MATHEMATICS *and*
DEMOCRACY

MATHEMATICS *and* DEMOCRACY

The Case for Quantitative Literacy

Prepared by

THE NATIONAL COUNCIL
ON EDUCATION AND THE DISCIPLINES

LYNN ARTHUR STEEN
Executive Editor

NCED

NATIONAL COUNCIL ON EDUCATION AND THE DISCIPLINES

The goal of the National Council on Education and the Disciplines (NCED) is to advance a vision that will unify and guide efforts to strengthen K–16 education in the United States. In pursuing this aim, NCED especially focuses on the continuity and quality of learning in the later years of high school and the early years of college. From its home at The Woodrow Wilson National Fellowship Foundation, NCED draws on the energy and expertise of scholars and educators in the disciplines to address the school–college continuum. At the heart of its work is a national reexamination of the core literacies—quantitative, scientific, historical, and communicative— that are essential to the coherent, forward-looking education all students deserve.

○

THE WOODROW WILSON NATIONAL FELLOWSHIP FOUNDATION

Founded in 1945, the Woodrow Wilson National Fellowship Foundation is an independent, nonprofit organization dedicated to the encouragement of excellence in education through the identification of critical needs and the development of effective national programs to address them. This combination of analysis and translation into action is the Foundation's unique contribution. Its programs include fellowships for graduate study, professional development for teachers, educational opportunities for women and minorities, relating the academy to society, and national service.

CONTENTS

CONTRIBUTORS

Design Team

The following contributors were members of the Design Team formed to look into the meaning of numeracy in contemporary society. Led by Lynn A. Steen, the Design Team developed the first part of this book, "The Case for Quantitative Literacy."

Gail Burrill is director of the Mathematical Sciences Education Board at the National Research Council in Washington, D.C.

Susan Ganter is associate professor of mathematical sciences in the Department of Mathematical Sciences at Clemson University in Clemson, South Carolina.

Daniel L. Goroff is professor of the practice of mathematics and associate director of the Derek Bok Center for Teaching and Learning at Harvard University.

Frederick P. Greenleaf is professor of mathematics in the Department of Mathematics at the Courant Institute of New York University in New York City.

W. Norton Grubb is David Gardner Professor of Higher Education Policy, Organization, Measurement, and Evaluation in the Graduate School of Education of the University of California at Berkeley.

Jerry Johnson is professor and chairman of the Mathematics Department at the University of Nevada at Reno.

Shirley M. Malcom is head of the Directorate for Education and Human Resources Programs at the American Association for the Advancement of Science in Washington, D.C.

Veronica Meeks teaches mathematics at Western Hills High School in Fort Worth, Texas.

Judith Moran is associate professor of quantitative studies and director of the Mathematics Center at Trinity College in Hartford, Connecticut.

Arnold Packer is chair of the SCANS 2000 Center at Johns Hopkins University in Baltimore, Maryland.

Janet P. Ray is a professor at Seattle Central Community College in Seattle, Washington.

C. J. Shroll is executive director of the Workforce Development Initiative at the Michigan Community College Association in Lansing.

Edward A. Silver is professor of mathematics in the Department of Mathematics in the University of Michigan School of Education in Ann Arbor.

Lynn Arthur Steen is professor of mathematics in the Department of Mathematics at St. Olaf College in Northfield, Minnesota.

Jessica Utts is a professor in the Department of Statistics at the University of California at Davis.

Dorothy Wallace is professor of mathematics in the Department of Mathematics at Dartmouth College in Hanover, New Hampshire.

Authors

The authors listed below were invited to comment on the case statement developed by the Quantitative Literacy Design Team.

Patricia Cline Cohen is professor of history at the University of California at Santa Barbara. Her primary area of research is nineteenth-century American history with emphasis on the history of women. Her recent publications include *A Calculating People: The Spread of Numeracy in Early America,* 1982, reissued 1999; *The Murder of Helen Jewett: The Life and Death of a Prostitute in Nineteenth-Century New York,* 1998, and "Numeracy in Nineteenth-Century America," in George Stanic (ed.), *A History of School Mathematics,* forthcoming.

Larry Cuban, professor of education at Stanford University, studies the history of school reform, including curriculum, governance, and technology. Among his recent publications is *Reconstructing the Common Good in Education: Coping with Intractable American Dilemmas* (coeditor with Dorothy Shipps), 2000.

Wade Ellis, Jr. is a mathematics instructor at West Valley College, in Saratoga, California. He is interested primarily in the use of technology in learning and teaching mathematics. His most recent publication is "Technology and Calculus" in Susan L. Ganter (ed.), *Calculus Renewal: Issues for Undergraduate Mathematics Education in the Next Decade,* 2000. He is coauthor of *Algebra I, Geometry, Algebra II,* 2000; *ODE Architect* (software package), 1999; *Calculus: Mathematics and Modeling,* 1998.

Peter T. Ewell is a senior associate at the National Center for Higher Education Management Systems (NCHEMS). He concentrates on the educational effectiveness of colleges and universities through applied research, policy development, and direct consulting with both institutions and state systems of higher education. His recent publications are "Achieving High Performance: The Policy Dimension" in W. G. Tierney (ed.), *The Responsive University,* 1998, and "Identifying Indicators of Curricular Quality" in J. G. Gaff and J. L. Ratcliff (eds.), *Handbook of the Undergraduate Curriculum,* 1996.

Deborah Hughes Hallett is professor of mathematics at the University of Arizona. Her area of research interest is undergraduate education in mathematics. She is coauthor of *Calculus* (with Andrew Gleason et al.), 2000; *Functions Modeling Change: A Preparation for Calculus* (with Eric Connally et al.), 2000, and "Teaching Quantitative Methods to Students of Public Affairs: Present and Future" in *Journal of Policy Analysis and Management,* Spring 2000.

Dan Kennedy is Lupton Professor of Mathematics at the Baylor School in Chattanooga, Tennessee. He is interested in all areas of education reform, particularly as they affect the transition from high school to college. He is coauthor of *Calculus: Graphical, Numerical, Algebraic* (Finney, Demana, Waits, Kennedy), 1999, and *Precalculus: Graphical, Numerical, Algebraic* (Demana, Waits, Foley, Kennedy), 2000.

Alfred B. Manaster is professor of mathematics at the University of California, San Diego. His professional activities are largely focused on the learning, teaching, and assessment of mathematics in courses from algebra through calculus. He serves as the director of California's Mathematics Diagnostic Testing Project and he was a member of one of the writing groups for NCTM's *Principles and Standards for School Mathematics,* 2000. His most recent research publication is "Some Characteristics of Eighth Grade Mathematics Classes in the TIMSS Videotape Study" in *The American Mathematical Monthly,* November 1998.

Joan L. Richards is associate professor in the History Department at Brown University. She is a historian of science with a primary interest in the development of mathematics and logic in the nineteenth century. Her publications include *Angles of Reflection: Logic and a Mother's Love,* 2000; "'In a Rational World All Radicals Would be Exterminated': Mathematics, Logic and Secular Thinking in Augustus De Morgan's England," *Science and Context,* in press, and "The Probable and the Possible in Early Victorian England" in Bernard Lightman (ed.), *Contexts of Victorian Science,* 1997.

Carol Geary Schneider is president of the Association of American Colleges and Universities (AAC&U). Her interests include an effort to rethink the broad aims of a twenty-first-century college education so that liberal learning becomes a framework for the entire educational experience. Her recent publications include "Core Missions and Civic Responsibility: Toward the Engaged Academy" in Thomas Ehrlich (ed.), *Civic Responsibility and Higher Education,* 2000; "From Diversity to Engaging Difference" in Nico Cloete and Mashew Miller (eds.), *Knowledge, Identity and Curriculum Transformation in Africa,* 1997. She is coauthor, with Robert Shoenberg, of "Contemporary Understandings of Liberal Education" in the AAC&U series on The Academy in Transition.

Alan H. Schoenfeld is the Elizabeth and Edward Conner Professor of Education at the University of California, Berkeley. His research is concerned with the nature of mathematical thinking, teaching, and learning. Schoenfeld was a writing group leader for *Principles and Standards for School Mathematics* and he has written, edited, or coedited books on mathematics education, including four volumes of *Research in Collegiate Mathematics Education* (1994, 1996, 1998, 2000); *Mathematical Thinking and Problem Solving,* 1994, and *Cognitive Science and Mathematics Education,* 1987.

James H. Stith is the director of the Physics Resources Center at the American Institute of Physics. His areas of interests are physics education (program evaluation and teacher preparation and enhancement) and student recruitment and retention. His recent publications include "Having a Private Conversation in a Crowded Room" in C. Stanley (ed.), *Large Class Instruction* (in press); "Getting Actively Involved in Professional Organizations" in S. M. McBay (ed.), *Scholarly Guideposts for Junior Faculty,* 2000, and "Interdisciplinary Communication and Understanding" with J. H. Grubbs in D. C. Arney (ed.), *Interdisciplinary Lively Application Projects* (ILAPS), 1997.

Zalman Usiskin is professor of education at the University of Chicago and director of the University of Chicago School Mathematics Project. His research primarily has involved the school mathematics curriculum, instruction, testing, and history and policy. He edited a special issue of *American Journal of Education,* "Reforming the Third R: Changing the School Mathematics Curriculum," November 1997. He was editor of *Developments in School Mathematics Around the World, Volume 4,* 1999, and his other publications include "The Development into the Mathematically Talented," *Journal of Secondary Gifted Education,* Spring 2000, and "Paper and Pencil Algorithms in a Calculator/Computer Age," in Lorna J. Morrow and Patricia J. Kenney (eds.), *Algorithms in School Mathematics,* 1998.

◐

Robert Orrill is the Executive Director, National Council on Education and the Disciplines (NCED), and Senior Advisor at The Woodrow Wilson National Fellowship Foundation (WWNFF), Princeton, New Jersey. NCED brings together university faculty and secondary school teachers to address issues of educational continuity in the later years of high school and the early years of college. Formerly Executive Director of the Office of Academic Affairs at the College Board, Mr. Orrill was chief liaison with the academic community, both at the secondary and postsecondary levels and across the full range of disciplines. Among his numerous publications are: *The Future of Education: Perspectives on National Standards in America,* 1994; *The Condition of American Liberal Education: Pragmatism and a Changing Tradition,* 1995, and *Education and Democracy: Re-imagining Liberal Learning in America,* 1997.

Lynn Arthur Steen is professor of mathematics and senior adviser to the academic vice president at St. Olaf College in Northfield, Minnesota. His current work is focused on numeracy—the quantitative and mathematical requirements for contemporary work and responsible citizenship. He is the editor or author of many books and articles on mathematics and education. Most recently he edited *Why Numbers Count: Quantitative Literacy for Tomorrow's America,* 1997. Earlier volumes include *On the Shoulders of Giants: New Approaches to Numeracy,* 1991; *Everybody Counts,* 1989; *Calculus for a New Century,* 1988.

Mathematics, Numeracy, and Democracy

In a retrospective account, the historian Lawrence Cremin cites rising rates of literacy as one of the most significant long-term achievements of American education. As a whole, Cremin argues, Americans were a more literate population at the end of the twentieth century than at its beginning or at any time earlier, and he attributes this outcome directly to increased public access to education. When evaluating the evidence, however, Cremin cautions against defining literacy as no more than rudimentary technical skills in reading, writing, and computing. Indeed, Cremin himself goes much further than this, saying that "if literacy [in the twentieth century] did involve the achievement of a technical skill, its meaning also depended on what an individual did with that technical skill, on how it was used, on what sort of material, with what frequency, and to what ends" (Cremin, 1988).

If we adopt Cremin's conception, literacy obviously becomes a complex rather than a simple matter. From this point of view, we regard it not merely as a measurable amount of technical skill, but also as a judgment about the nature and quality of an interaction between a person with that skill and a particular social or environmental situation. This implies, in turn, that literacy can have no permanent meaning, no definition forever fixed and constant across all times and places. As we know well, social and environmental situations change, often because of human intervention, and thus what counts as literacy very likely will vary at least somewhat in different historical periods and from one cultural setting to another. This is a recognition that Cremin's mentor, John

Dewey, unceasingly urged upon educators and the American public generally. Under modern American conditions, Dewey said, change repeatedly outruns continuity, and thus the need to reframe the meaning of literacy must continue apace through a process that "can never be ended." Such, he observed, "is the need of a human nature and of a society that are themselves in process of constant change" (Dewey, 1931).

For both Dewey and Cremin, the matter becomes even more complex when we ask what literacy means in a society dedicated to democratic ideals and informed by an ethos of individual freedom. In democratic settings, Cremin says, it is important to distinguish between what he calls "inert" and "liberating" literacy. As Cremin defines these terms, the former is that level of verbal and numerate skill required to comprehend instructions, perform routine procedures, and complete tasks in a rote manner. From a social perspective, this is that measure of literacy we might expect to find applied in a cultural setting in which tradition prevails and customs are securely in place, and where opportunities for freedom, choice, and innovation are limited. To speak of literacy as "liberating," however, assumes a much more challenging standard by which individuals command both the enabling skills needed to search out information and the power of mind necessary to critique it, reflect upon it, and apply it in making decisions. It is only this more expansive and demanding meaning of literacy, or what Dewey calls "popular enlightenment," that can inform and animate a vital democracy. Indeed, Dewey reminds us, a successful democracy is conceivable only when and where individuals are able to "think for themselves," "judge independently," and discriminate between good and bad information.

Turning to the present, the considerations raised by Dewey and Cremin bear directly on the issues addressed in this book. Currently, to be sure, the optimism that colors their views has been replaced by an attitude of concern, or even foreboding, in some quarters. Indeed, many no longer believe that literacy is as prevalent in American society as it once was. Of these, the most outspoken contend that literacy rates are on the decline in this country and that the long-standing historical trend lauded by Cremin is in serious danger of being reversed. As further cause for alarm, they assert, there is evidence that American students do poorly on tests of literacy skills when compared with the performance of students from many other nations—thus raising the specter of global competitive disadvantage along with that of social dysfunction at home (Hirsch, 1996).

Others question these dark claims and say that there is no good evidence to support them, but few, if any, dispute that we can and should do better in making efforts to nurture literacy throughout society. This consensus has helped move literacy to the top of the national social agenda, and it now appears highly probable that both of the major political parties will present ambitious new proposals aimed at strengthening literacy through education, testing, and other means. Whatever the wisdom of these plans, however, we can be sure that each will invoke the belief still deeply held by most Americans that literacy is a great enabler—that, as Dewey said, it is the necessary prerequisite for access to "a life of widened freedom" and the competence needed for each and every individual to have the opportunity to be all he or she is capable of becoming.

This attention to literacy is very welcome and much needed, but arriving at sensible approaches to the educational issues involved will require asking what we mean when we speak of a literate person. Looking forward, what should be our conception of literacy and what are our standards for its achievement at the advent of a new century? In a time of uncertainty about educational achievement, such as now exists, there can be a disposition to answer this question by hearkening back to the ideas and standards of some previous era. Because they are easier to pin down, these convictions of an earlier time may seem to provide an element of constancy in the midst of shifting conditions. But, as Dewey insists, how can thinking about literacy look back, and in effect stand still or even regress, when "society itself is changing under our very eyes"? In practice, would not this result in *mis*education that fails to take account of the very forces propelling change that we most need to understand and guide?

Today, this question is particularly pressing with regard to numeracy—the analogue to reading and writing in the triumvirate of competencies that, conjoined and working in unison, make up the traditional core of literacy. Few would disagree that, with the arrival of the computer age, the environing conditions that must be addressed in arriving at a definition of numeracy are undergoing rapid and often bewildering change. Increasingly, as the contributors to this book note, we live in a society "awash in numbers" and "drenched with data." Throughout most of history, human beings had to make do with sparse and incomplete information about the world because typically data were difficult to obtain and insufficient to the task. Now, however, the flood of available information produced by powerful computers and their many applications threatens to become an overwhelming deluge. Working nonstop and with extraordinary speed, computers meticulously and relentlessly note details about the world around them and carefully record these details. As a result, they create data in increasing amounts every time a purchase is made, a poll is taken, a disease is diagnosed, or a satellite passes over a section of terrain. In consequence, one observer notes, whereas until very recently information about the world has been scarce and hard to come by, "today we are drowning in data, and there is unimaginably more on the way" (Bailey, 1996).

The implications of this new situation can be either very good or very bad. At present, they are some of both. For those competent and comfortable in thinking with numbers, the opportunities that come with the new conditions can be liberating. Not only specialists but now everyone can obtain and consider data about the risks of medication, voting patterns in a locality, projections for the federal budget surplus, and an almost endless array of other concerns. Potentially, if put to good use, this unprecedented access to numerical information promises to place more power in the hands of individuals and serve as a stimulus to democratic discourse and civic decision making. Indeed, as a recent study illustrates, the availability of numbers now reaches into "every nook and cranny of American life," making it no exaggeration to say that, in consequence, numerical thinking has become essential to "the discourse of public life" (Caplow et al., 2001). It follows, however, that if individuals lack the ability to think numerically they cannot participate fully in civic life, thereby bringing into question the very basis of government of, by, and for the people.

Moreover, the consequences of what John Allen Paulos named "innumeracy" (Paulos, 1988) can be profoundly disabling in every sphere of human endeavor—whether it be in home and private life, work and career, or public and professional pursuits. Stating the case in dramatic terms, Lynn Steen warns that "an innumerate citizen today is as vulnerable as the illiterate peasant of Gutenberg's time" (Steen, 1997). Any such possibility of regress to pre-Enlightenment conditions would be deeply troubling under any circumstances and most certainly is unacceptable in a democracy. But how can we address the danger? What actions should we take? To begin, what exactly is quantitative literacy in today's world? How do we define and describe it in such a way that steps can be taken to foster it? In search of answers, the National Council on Education and the Disciplines (NCED) put these questions to a Quantitative Literacy Design Team formed for the purpose of inquiring into the meaning of numeracy in contemporary society. Led by Steen, the Design Team chose to cast its reply in the form of the case statement now made available for the first time in this book. As readers will see, the result is a rounded and thoroughgoing consideration of that way of thinking about the world we have come to call numeracy.

That said, however, the case statement does not seek to end debate about the meaning of numeracy. On the contrary, it aspires instead to be a starting point for a much needed wider conversation. Most certainly, this conversation must be carried forward first and foremost in school and college settings. If asked, faculty and administrators at most schools and colleges today probably would say that they intend to produce numerate graduates, although they might use different vocabularies to describe this aim. At the same time, however, if we look closely it is difficult to identify very many academic institutions at which extensive consideration has been given to the meaning of this outcome or to how students pursue and achieve it. We believe that the case statement can be a helpful point of departure for just such a consideration, and, with this in mind, NCED invited 12 respondents to comment on the views expressed by the Design Team. Like the members of the Design Team, each of the respondents brings a wealth of experience and insight to quantitative issues. Their responses make up the second half of this publication, and each adds to and carries forward the conversation from a different perspective. Together with the case statement, they provide an excellent beginning for a national discussion of the increasingly important links among mathematics, numeracy, and democracy in the changing world of the twenty-first century.

We hasten to add, however, that this conversation is not for educators alone. In every way possible, the public must be encouraged to join the discussion. As Steen points out in the epilogue to this book, many commonly held assumptions about the relationship of mathematics and quantitative literacy impede understanding and are therefore much in need of reexamination. It may seem only common sense, Steen says, that a rigorous education in mathematics along traditional lines should lead to a high degree of achievement in quantitative literacy. Contrary to this popular belief, however, only a small part of the education needed to attain control over numbers can be found in the typical mathematics curriculum. That is because skills in complex counting and data analysis, like many other aspects of numeracy described in the case statement, rarely find a place in the standard calculus-oriented mathematics progression. Once basic arithmetic is left behind early in a student's education, the mathematics curriculum moves on to more abstract concepts that are most applicable for future work in a limited number of technical professions. Only to a limited extent do students engage in the kind of quantitative work needed in the great variety of contexts and settings that they will encounter in life.

An important theme of this volume, then, is that efforts to intensify attention to the traditional mathematics curriculum do not necessarily lead to increased competency with quantitative data and numbers. While perhaps surprising to many in the public, this conclusion follows from a simple recognition—that is, unlike mathematics, numeracy does not so much lead upward in an ascending pursuit of abstraction as it moves outward toward an ever richer engagement with life's diverse contexts and situations. When a professional mathematician is most fully at work, Keith Devlin writes, the process becomes so abstract and inward that "the mathematician must completely shut out the outside world" (Devlin, 2000). The numerate individual, by contrast, seeks out the world and uses quantitative skills to come to grips with its varied settings and concrete particularity.

This is not to say, of course, that mathematics and numeracy have little to do with one another in a balanced education. There is a sense, in fact, in which numeracy should be thought of as the extension of mathematics into other subjects in which, too often, the quantitative aspects of life are ignored altogether. This kind of compartmentalization, however, is rarely possible in the world outside. In life, numbers are everywhere and cannot be segregated into one subject and left out of others as often happens when we build our academic cubbyholes. Indeed, if the need for quantitative competence is now pervasive in American life in the many ways this volume indicates, it seems only common sense that the responsibility for fostering quantitative literacy should be spread broadly across the curriculum. This book is a step toward bringing attention to the compelling arguments for arriving at this conclusion and, as it should follow, to making quantitative thought much more than an affair of the mathematics classroom alone.

Readers, therefore, should not come to this book expecting only to hear arguments about at what age students should begin to learn algebra or whether all should aspire to study calculus. Though not unimportant issues, there is something missing in these debates when we consider them in light of the quantitative demands of contemporary life. In fact, it may be the most significant question that is not being asked. One way to approach the debate is to assume the traditional mathematics curriculum and ask how more students can succeed in it. Another way is to consider the quantitative challenges that arise day in and day out in American life and ask what kind of education would lead to the fully liberating literacy that Dewey and Cremin say we must seek in a democracy. While neither is the one right and only approach, this book takes the latter course, and, in so doing, the contributors hope to launch a conversation that is widespread, rich, and productive of new thinking about the educational outcomes we most want for students.

◑

The making of a book is a cooperative undertaking. Readers will find ample evidence of this in the list of contributors. Among the many others who helped, very special thanks must go to Lynn Steen, who guided the project from start to finish. Our thanks also go to Diane Foster, whose superb editorial skills and all-around good sense entered into every decision made from the time the idea of a book first came into our minds. When the book was in draft, Susan Ganter's close reading of each and every contribution resulted in many improvements, and Dorothy Downie made sure that we attended to details large and small as she so often has done in the past. Finally, Mary Catherine Snyder will know what I mean in mentioning that she saved the day more than once.

In closing, let me also thank Bob Weisbuch and the staff of The Woodrow Wilson National Fellowship Foundation for providing the supportive work environment in which a book of this kind could be

produced. The book itself was made possible by financial assistance provided by the Pew Charitable Trusts, and I am very pleased to have this
opportunity to thank the Trusts for the support given to the National
Council on Education and the Disciplines.

Robert Orrill

EXECUTIVE DIRECTOR
NATIONAL COUNCIL ON
EDUCATION AND THE DISCIPLINES

REFERENCES

Bailey, James. *After Thought: The Computer Challenge to Human Intelligence.*
New York, NY: Basic Books, 1996.

Caplow, Theodore, Louis Hicks, and Ben J. Wattenberg. *The First Measured
Century: An Illustrated Guide to Trends in America, 1900–2000.* Washington,
DC: AEI Press, 2001.

Cremin, Lawrence A. *American Education: The Metropolitan Experience
1876–1980.* New York, NY: Harper & Row, 1988.

Devlin, Keith. *The Math Gene: How Mathematical Thinking Evolved and Why
Numbers Are Like Gossip.* New York, NY: Basic Books, 2000.

Dewey, John. "American Education Past and Future," *The Later Works,* volume
6:1931–32. Jo Ann Boydston (Editor), Southern Illinois University Press.

Hirsch, E. D., Jr. *The Schools We Need and Why We Don't Have Them.* New
York, NY: Doubleday, 1996.

Paulos, John Allen. *Innumeracy: Mathematical Illiteracy and Its Consequences.*
New York, NY: Vintage Books, 1988.

Steen, Lynn Arthur (Editor), *Why Numbers Count: Quantitative Literacy for
Tomorrow's America.* New York, NY: The College Board, 1997.

THE CASE FOR QUANTITATIVE LITERACY

The world of the twenty-first century is a world awash in numbers. Headlines use quantitative measures to report increases in gasoline prices, changes in SAT scores, risks of dying from colon cancer, and numbers of refugees from the latest ethnic war. Advertisements use numbers to compete over costs of cell phone contracts and low-interest car loans. Sports reporting abounds in team statistics and odds on forthcoming competitions.

More important for many people are the rapidly increasing uses of quantitative thinking in the workplace, in education, and in nearly every other field of human endeavor. Farmers use computers to find markets, analyze soil, and deliver controlled amounts of seeds and nutrients; nurses use unit conversions to verify accuracy of drug dosages; sociologists draw inferences from data to understand human behavior; biologists develop computer algorithms to map the human genome; factory supervisors use "six-sigma" strategies to ensure quality control; entrepreneurs project markets and costs using computer spreadsheets; lawyers use statistical evidence and arguments involving probabilities to convince jurors. The roles played by numbers and data in contemporary society are virtually endless.

Unfortunately, despite years of study and life experience in an environment immersed in data, many educated adults remain functionally innumerate. Most U.S. students leave high school with quantitative skills far below what they need to live well in today's society; businesses lament the lack of technical and quantitative skills of prospective employees; and virtually every college finds that many students need remedial mathematics. Data from the National Assessment of Educational Progress (NAEP) show that the average mathematics performance of seventeen-year-old students has risen just one percent in 25 years and remains, at 307, in the lower half of the "basic" range (286–336) and well below the "proficient" range (336–367). Moreover, despite slight growth in recent years, average scores of Hispanic students (292) and black students (286) are near the bottom of the "basic" range (NCES, 1997).

Common responses to this well-known problem are either to demand more years of high school mathematics or more rigorous standards for graduation. Yet even individuals who have studied trigonometry and calculus often remain largely ignorant of common abuses of data and all too often find themselves unable to comprehend (much less to articulate) the nuances of quantitative inferences. As it turns out, it is not calculus but numeracy that is the key to understanding our data-drenched society.

Quantitatively literate citizens need to know more than formulas and equations. They need a predisposition to look at the world through mathematical eyes, to see the benefits (and risks) of thinking quantitatively about commonplace issues, and to approach complex problems with confidence in the value of careful reasoning. Quantitative literacy empowers people by giving them tools to think for themselves, to ask intelligent questions of experts, and to confront authority confidently. These are skills required to thrive in the modern world.

A Brief History of Quantitative Literacy

Although the discipline of mathematics has a very ancient history—both as a logical system of axioms, hypotheses, and deductions and as a tool for empirical analysis of the natural world—the expectation that ordinary citizens be quantitatively literate is primarily a phenomenon of the late twentieth century. In ancient times, numbers, especially large numbers, served more as metaphors than as measurements. The importance of quantitative methods in the lives of ordinary people emerged very slowly in the middle ages as artists and merchants learned the value of imposing standardized measures of length, time, and money on their arts and crafts—for example, in polyphonic music, perspective drawing, and double-entry bookkeeping (Crosby, 1997).

In colonial America, leaders such as Franklin and Jefferson promoted numeracy to support the new experiment in popular democracy, even as skeptics questioned the legitimacy of policy arguments based on empirical rather than religious grounds (Cohen, 1982). Only in the latter part of the twentieth century did quantitative methods achieve their current status as the dominant form of acceptable evidence in most areas of public life (Bernstein, 1996; Porter, 1995; Wise, 1995). Despite their origins in astrology, numerology, and eschatology, numbers have become the chief instruments through which we attempt to exercise control over nature, over risk, and over life itself.

As the gap has widened between the quantitative needs of citizens and the quantitative capacity of individuals, publications about "math anxiety" and "math panic" have raised public awareness of the consequences of innumeracy (Buxton, 1991; Paulos, 1988, 1996; Tobias, 1978, 1993). At the same time, publications such as Edward Tufte's extraordinary volumes on the visual display of quantitative information reveal the unprecedented power of quantitative information to communicate and persuade (Tufte, 1983, 1990, 1997). We see the results every day, both good and bad, in the widespread practice in newspapers of using charts and graphs as the preferred means of presenting quantitative information.

In 1989 the National Council of Teachers of Mathematics (NCTM) responded to the changing mathematical needs of society by publishing standards for school mathematics that called for all students to learn rich and challenging mathematics. Subsequently, other standards documented the role of quantitative methods in education (e.g., science, history, geography, social studies) and careers (e.g., bioscience, electronics, health care, photonics). In April 2000 NCTM released a much-anticipated update of its standards for school mathematics (NCTM, 2000). These standards and their interpretations in state frameworks, textbooks, curricula, and assessments have engendered considerable public debate about the goals of education and about the relation of mathematics to these goals.

In recognition of the increasing importance of quantitative literacy in the lives of nations, government agencies that monitor literacy divided what had been a single concept into three components: prose, document, and quantitative literacy (Kirsch and Jungeblut, 1986; NCES, 1993; OECD, 1995, 1998). Similar awareness led many liberal arts colleges to infuse quantitative methods into courses in the arts and humanities (White, 1981). At the same time, economists expanded the traditional "3 R's" requirement for employment (reading, 'riting, 'rithmetic) to encompass five additional competencies: resources, interpersonal, information, systems, and technology (SCANS, 1991). More recent publications have examined the role of quantitative literacy in relation to the changing economy (Murnane and Levy, 1996), the expectations of college graduates (Sons, 1996), the perspectives of professionals in a variety of fields (Steen, 1997), and the demands of the high-performance workplace (Forman and Steen, 1999).

The footprints of quantitative literacy can be found throughout these publications, but not clarity about its meaning. These sources reveal more confusion than consensus about the nature of quantitative literacy, especially about its relation to mathematics. They echo the historical dichotomy of mathematics as academic and numeracy as commercial, and pay at most lip service to the role numeracy plays in informing citizens and supporting democratic government. What we learn is that although almost everyone believes quantitative literacy to be important, there is little agreement on just what it is.

Mathematics, Statistics, and Quantitative Literacy

In the beginning, grammar schools taught arithmetic and colleges, mathematics. As secondary schools became the transition from grammar school to college, courses in algebra, geometry, trigonometry, analytic geometry, and even calculus created a highway that led increasing numbers of students directly from arithmetic to higher mathematics. At the same time, mathematics itself expanded into a collection of mathematical sciences that now includes, in addition to traditional pure and applied mathematics, subjects such as statistics, financial mathematics, theoretical computer science, operations research (the science of optimization), and newest of all, bioinformatics. Although each of these subjects shares with mathematics many foundational tools, each has its own distinctive character, methodologies, standards, and accomplishments.

The mathematical science that ordinary individuals most often encounter is statistics, originally the science of the state (as in census). Statistics underlies every clinical trial, every opinion survey, and every government economic report. Yet school curricula still primarily serve to prepare students only for traditional college mathematics. School mathematics places relatively little emphasis on topics designed to build a bridge from arithmetic to the subtle and fascinating world of statistics. Recognizing this neglect, the American Statistical Association (ASA) and the NCTM have cooperated for many years in a campaign to infuse more exploratory data analysis and elementary statistics into school curricula. This effort, interestingly, is called the "Quantitative Literacy Project." (Project founders chose quantitative literacy rather than statistics as a title because they anticipated public anxiety about the term *statistics*.)

Despite its occasional use as a euphemism for statistics in school curricula, quantitative literacy is not the same as statistics. Neither is it the same as mathematics, nor is it (as some fear) watered-down mathematics. Quantitative literacy is more a habit of mind, an approach to problems that employs and enhances both statistics and mathematics. Unlike statistics, which is primarily about uncertainty, numeracy is often about the logic of certainty. Unlike mathematics, which is primarily about a Platonic realm of abstract structures, numeracy is often anchored in data derived from and attached to the empirical world. Surprisingly to some, this inextricable link to reality makes quantitative reasoning every bit as challenging and rigorous as mathematical reasoning. (Indeed, evidence from Advanced Placement examinations suggests that students of comparable ability find data-based statistical reasoning more difficult than symbol-based mathematical reasoning.)

Connecting mathematics to authentic contexts demands delicate balance. On the one hand, contextual details camouflage broad patterns that are the essence of mathematics; on the other hand, these same details offer associations that are critically important for many students' long-term learning. Few can doubt that the tradition of decontextualized mathematics instruction has failed many students, including large numbers of women and minorities, who leave high school with neither the numeracy skills nor the quantitative confidence required in contemporary society. The tradition of using mathematics as a filter for future academic performance is reinforced by increasing demand for admission to selective colleges and universities. These pressures skew school curricula in directions that are difficult to justify because they leave many students functionally innumerate.

Whereas the mathematics curriculum has historically focused on school-based knowledge, quantitative literacy involves mathematics acting in the world. Typical numeracy challenges involve real data and uncertain procedures but require primarily elementary mathematics. In contrast, typical school mathematics problems involve simplified numbers and straightforward procedures but require sophisticated abstract concepts. The test of numeracy, as of any literacy, is whether a person naturally uses appropriate skills in many different contexts.

Educators know all too well the common phenomenon of compartmentalization, when skills or ideas learned in one class are totally forgotten when they arise in a different context. This is an especially acute problem for school mathematics, in which the disconnect from meaningful contexts creates in many students a stunning absence of common number sense. To be useful for the student, numeracy needs to be learned and used in multiple contexts—in history and geography, in economics and biology, in agriculture and culinary arts (Steen, 1998, 2000). Numeracy is not just one among many subjects but an integral part of all subjects.

Elements of Quantitative Literacy

The capacity to deal effectively with the quantitative aspects of life is referred to by many different names, among them quantitative literacy, numeracy, mathematical literacy, quantitative reasoning, or sometimes just plain "mathematics." Different terms, however, convey different nuances and connotations that are not necessarily interpreted in the same way by all listeners.

An early definition of the term *numerate,* widely cited by mathematics educators, appeared in a British government report on mathematics education (Cockcroft, 1982):

> We would wish the word numerate to imply the possession of two attributes. The first of these is an "at homeness" with numbers and an ability to make use of mathematical skills which enables an individual to cope with the practical demands of everyday life. The second is an ability to

have some appreciation and understanding of information which is presented in mathematical terms.

The same two themes emerged in the National Adult Literacy Survey (NCES, 1993), which defined *quantitative literacy* as:

> The knowledge and skills required to apply arithmetic operations, either alone or sequentially, using numbers embedded in printed material (e.g., balancing a checkbook, completing an order form).

The National Center for Education Statistics (NCES) defines the closely related knowledge and skills required to locate and use information (for example, in payroll forms, transportation schedules, maps, tables, and graphs) as *document literacy*. In contrast, the International Life Skills Survey (ILSS, 2000) currently underway defines *quantitative literacy* in a much more comprehensive manner as:

> An aggregate of skills, knowledge, beliefs, dispositions, habits of mind, communication capabilities, and problem solving skills that people need in order to engage effectively in quantitative situations arising in life and work.

The Programme for International Student Assessment (PISA, 2000) adopts a similar definition but calls it *mathematics literacy:*

> An individual's capacity to identify and understand the role that mathematics plays in the world, to make well-founded mathematical judgements and to engage in mathematics in ways that meet the needs of that individual's current and future life as a constructive, concerned and reflective citizen.

From just these four definitions significant differences emerge. Some focus on an individual's ability to use quantitative tools, others on the ability to understand and appreciate the role of mathematical and quantitative methods in world affairs. Some emphasize basic skills ("arithmetic operations"), others higher-order thinking ("well-founded judgements"). To clarify these different definitions, as well as to make them more useful, we break them into different elements, which may be combined, as atoms in molecules, to form a more comprehensive portrait of quantitative literacy. Here are some of these elements:

Confidence with Mathematics. Being comfortable with quantitative ideas and at ease in applying quantitative methods. Individuals who are quantitatively confident routinely use mental estimates to quantify, interpret, and check other information. Confidence is the opposite of "math anxiety"; it makes numeracy as natural as ordinary language.

Cultural Appreciation. Understanding the nature and history of mathematics, its role in scientific inquiry and technological progress, and its importance for comprehending issues in the public realm.

Interpreting Data. Reasoning with data, reading graphs, drawing inferences, and recognizing sources of error. This perspective differs from traditional mathematics in that data (rather than formulas or relationships) are at the center.

Logical Thinking. Analyzing evidence, reasoning carefully, understanding arguments, questioning assumptions, detecting fallacies, and evaluating risks. Individuals with such habits of inquiry accept little at face value; they constantly look beneath the surface, demanding appropriate information to get at the essence of issues.

Making Decisions. Using mathematics to make decisions and solve problems in everyday life. For individuals who have acquired this habit, mathematics is not something done only in mathematics class but a powerful tool for living, as useful and ingrained as reading and speaking.

Mathematics in Context. Using mathematical tools in specific settings where the context provides meaning. Notation, problem-solving strategies, and performance standards all depend on the specific context.

Number Sense. Having accurate intuition about the meaning of numbers, confidence in estimation, and common sense in employing numbers as a measure of things.

Practical Skills. Knowing how to solve quantitative problems that a person is likely to encounter at home or at work. Individuals who possess these skills are adept at using elementary mathematics in a wide variety of common situations.

Prerequisite Knowledge. Having the ability to use a wide range of algebraic, geometric, and statistical tools that are required in many fields of postsecondary education.

Symbol Sense. Being comfortable using algebraic symbols and at ease in reading and interpreting them, and exhibiting good sense about the syntax and grammar of mathematical symbols.

These elements illuminate but do not resolve the linguistic confusions that permeate discussions of quantitative literacy. Sometimes the terms *quantitative* and *mathematical* are used interchangeably, but often they are used to signify important distinctions—for example, between what is needed for life (quantitative) and what is needed for education (mathematics), or between what is needed for general school subjects (quantitative) and what is needed for engineering and physical science (mathematics). For some the word *quantitative* seems too limiting, suggesting numbers and calculation rather than reasoning and logic, while for others the term seems too vague, suggesting a diminution of emphasis on traditional mathematics. Similarly, the term *literacy* conveys different meanings: for some it suggests a minimal capacity to read, write, and calculate, while for others it connotes the defining characteristics of an educated (literate) person.

In terms of what is needed for active and alert participation in contemporary society, quantitative literacy can be viewed as a direct analog of verbal literacy. At a fundamental level we teach the skills of reading, writing, and calculating, the principal goals of lower schools. But these basic skills are no longer sufficient to sustain a successful career or to participate fully in a modern democratic society. Today's well-educated citizens require sophistication in both literacy and numeracy to think through subtle issues that are communicated in a collage of verbal, symbolic, and graphic forms. In addition, they need the confidence to express themselves in any of these modern forms of communication. In the twenty-first century, literacy and numeracy will become inseparable qualities of an educated person.

Expressions of Quantitative Literacy

A different way to think about quantitative literacy is to look not at definitions but at actions, not at what numeracy is but at how it is expressed. Many manifestations are commonplace and obviously important, yet they are not the real reason for the increasing emphasis on numeracy.

Examples:

- Estimating how to split a lunch bill three ways
- Comparing price options for leasing or purchasing a car
- Reading and understanding nutrition labels
- Reconciling a bank statement and locating the sources of error
- Scaling recipes up and down and converting units of volume and weight
- Mentally estimating discounts, tips, and sales prices
- Understanding the effects of compound interest
- Reading bus schedules and maps

More relevant to current students and future citizens are many of the more sophisticated expressions of quantitative reasoning that have become common in our data-driven society. Some of these serve primarily personal ends, while others serve the goals of a democratic society. Together they provide a rich portrait of numeracy in the modern world.

Citizenship

Virtually every major public issue—from health care to social security, from international economics to welfare reform—depends on data, projections, inferences, and the kind of systematic thinking that is at the heart of quantitative literacy. Examples:

- Understanding how resampling and statistical estimates can improve the accuracy of a census
- Understanding how different voting procedures (e.g., runoff, approval, plurality, preferential) can influence the results of elections
- Understanding comparative magnitudes of risk and the significance of very small numbers (e.g., 10 ppm or 250 ppb)
- Understanding that unusual events (such as cancer clusters) can easily occur by chance alone
- Analyzing economic and demographic data to support or oppose policy proposals
- Understanding the difference between rates and changes in rates, for example, a decline in prices compared with a decline in the rate of growth of prices

- Understanding the behavior of weighted averages used in ranking colleges, cities, products, investments, and sports teams
- Appreciating common sources of bias in surveys such as poor wording of questions, volunteer response, and socially desirable answers
- Understanding how small samples can accurately predict public opinion, how sampling errors can limit reliability, and how sampling bias can influence results
- Recognizing how apparent bias in hiring or promotion may be an artifact of how data are aggregated
- Understanding quantitative arguments made in voter information pamphlets (e.g., about school budgets or tax proposals)
- Understanding student test results given in percentages and percentiles and interpreting what these data mean with respect to the quality of schools

Culture

As educated men and women are expected to know something of history, literature, and art, so should they know—at least in general terms—something of the history, nature, and role of mathematics in human culture. This aspect of quantitative literacy is most commonly articulated in goals colleges set forth for liberal education. Examples:

- Understanding that mathematics is a deductive discipline in which conclusions are true only if assumptions are satisfied
- Understanding the role mathematics played in the scientific revolution and the roles it continues to play
- Understanding the difference between deductive, scientific, and statistical inference
- Recognizing the power (and danger) of numbers in shaping policy in contemporary society
- Understanding the historical significance of zero and place value in our number system
- Knowing how the history of mathematics relates to the development of culture and society
- Understanding how assumptions influence the behavior of mathematical models and how to use models to make decisions

Education

Fields such as physics, economics, and engineering have always required a strong preparation in calculus. Today, other aspects of quantitative literacy (e.g., statistics and discrete mathematics) are also important in these fields. Increasingly, however, other academic disciplines are requiring that students have significant quantitative preparation. Examples:

- Biology requires computer mathematics (for mapping genomes), statistics (for assessing laboratory experiments), probability (for studying heredity), and calculus (for determining rates of change).
- Medicine requires subtle understanding of statistics (to assess clinical trials), of chance (to compare risks), and of calculus (to understand the body's electrical, biochemical, and cardiovascular systems).
- The social sciences rely increasingly on data either from surveys and censuses or from historical or archeological records; thus statistics is as important for a social science student as calculus is for an engineering student.
- Advances in scientific understanding of the brain have transformed psychology into a biological science requiring broad understanding of statistics, computer science, and other aspects of quantitative literacy.
- The stunning impact of computer graphics in the visual arts (film, photography, sculpture) has made parts of mathematics, especially calculus, geometry, and computer algorithms, very important in a field that formerly was relatively unquantitative.
- Interpretation of historical events increasingly depends on analysis of evidence provided either by numerical data (e.g., government statistics, economic indicators) or through verification and dating of artifacts.
- Even the study of language has been influenced by quantitative and logical methods, especially in linguistics, concordances, and the new field of computer translation.

Professions

As interpretation of evidence has become increasingly important in decisions that affect people's lives, professionals in virtually every field are now expected to be well versed in quantitative tools. Examples:

- Lawyers rely on careful logic to build their cases and on subtle arguments about probability to establish or refute "reasonable doubt."
- Doctors need both understanding of statistical evidence and the ability to explain risks with sufficient clarity to ensure "informed consent."
- Social workers need to understand complex state and federal regulations about income and expenses to explain and verify their clients' personal budgets.
- School administrators deal regularly with complex issues of scheduling, budgeting, inventory, and planning—all of which have many quantitative dimensions.
- Journalists need a sophisticated understanding of quantitative issues (especially of risks, rates, samples, surveys, and statistical evidence) to develop an informed and skeptical understanding of events in the news.
- Chefs use quantitative tools to plan schedules, balance costs against value of ingredients, and monitor nutritional balance of meals.
- Architects use geometry and computer graphics to design structures, statistics and probability to model usage, and calculus to understand engineering principles.

Personal Finance

Managing money well is probably the most common context in which ordinary people are faced with sophisticated quantitative issues. It is also an area greatly neglected in the traditional academic track of the mathematics curriculum. Examples:

- Understanding depreciation and its effect on the purchase of cars or computer equipment
- Comparing credit card offers with different interest rates for different periods of time
- Understanding the relation of risk to return in retirement investments
- Understanding the investment benefits of diversification and income averaging
- Calculating income tax and understanding the tax implications of financial decisions
- Estimating the long-term costs of making lower monthly credit card payments

- Understanding interactions among different factors affecting a mortgage (e.g., principal, points, fixed or variable interest, monthly payment, and duration)
- Using the Internet to make decisions about travel plans (routes, reservations)
- Understanding that there are no schemes for winning lotteries
- Choosing insurance plans, retirement plans, or finance plans for buying a house

Personal Health

As patients have become partners with doctors in making decisions about health care and as medical services have become more expensive, quantitative skills have become increasingly necessary in this important aspect of people's lives. Examples:

- Interpreting medical statistics and formulating relevant questions about different options for treatment in relation to known risks and the specifics of a person's condition
- Understanding medical dosages in relation to body weight, timing of medication, and drug interactions
- Weighing costs, benefits, and health risks of heavily advertised new drugs
- Understanding terms and conditions of different health insurance policies; verifying accuracy of bills and insurance payments
- Calibrating eating and exercise habits in relation to health
- Understanding the impact of outliers on summaries of medical data

Management

Many people need quantitative skills to manage small businesses or non-profit organizations as well as to fulfill their responsibilities when they serve on boards or committees that are engaged in running any kind of enterprise. Examples:

- Looking for patterns in data to identify trends in costs, sales, and demand
- Developing a business plan, including pricing, inventory, and staffing strategies for a small retail store

- Determining the break-even point for manufacturing and sale of a new product
- Gathering and analyzing data to improve profits
- Reviewing the budget of a small nonprofit organization and understanding relevant trends
- Understanding the limitations of extrapolating from data in a fixed range
- Calculating time differences and currency exchanges in different countries

Work

Virtually everyone uses quantitative tools in some way in relation to their work, if only to calculate their wages and benefits. Many examples of numeracy on the job are very specific to the particular work environment, but some are not. Examples:

- Producing a schedule or tree diagram for a complicated project
- Researching, interpreting, and using work-related formulas
- Using spreadsheets to model different scenarios for product sales and preparing graphs that illustrate these options
- Understanding and using exponential notation and logarithmic scales of measurement
- Maintaining and using quality control charts
- Optimizing networks to develop efficient ways to plan work processes
- Understanding the value of statistical quality control and statistical process control

Skills of Quantitative Literacy

For a different and more traditional perspective on quantitative literacy, we might create an inventory of quantitative skills expected of an educated person in contemporary society. For many, a list of skills is more comforting than a list of elements or expressions because skills are more immediately recognizable as something taught and learned in school. Moreover, many people believe that skills must precede applications and that once learned, quantitative skills can be applied whenever needed. Unfortunately, considerable evidence about the associative nature of learning suggests that this approach works very imperfectly. For most students, skills learned free of context are skills devoid of meaning and utility. To be effective, numeracy skills must be taught and learned in settings that are both meaningful and memorable.

Nevertheless, a list of skills is a valuable enhancement to our emerging definition of quantitative literacy—a third dimension, so to speak, which complements the foregoing analyses in terms of elements and expressions. A list of skills helps instructors plan curricula to cover important topics and helps examiners assess the desired balance of knowledge. An appendix to the Mathematical Association of America's report on quantitative literacy (Sons, 1996) offers—with suitable apologies and caveats—a consensus among mathematicians on skills that are especially important for courses in quantitative literacy. This list includes predictable topics from arithmetic, geometry, and algebra that are part of every school mathematics program, but it also includes many newer topics from statistics and optimization that are usually offered to students, if at all, only as electives.

In fact, many of these "elective" skills are firmly embedded in the elements and expressions of quantitative literacy. They include:

- *Arithmetic:* Having facility with simple mental arithmetic; estimating arithmetic calculations; reasoning with proportions; counting by indirection (combinatorics).
- *Data:* Using information conveyed as data, graphs, and charts; drawing inferences from data; recognizing disaggregation as a factor in interpreting data.
- *Computers:* Using spreadsheets, recording data, performing calculations, creating graphic displays, extrapolating, fitting lines or curves to data.
- *Modeling:* Formulating problems, seeking patterns, and drawing conclusions; recognizing interactions in complex systems; understanding linear, exponential, multivariate, and simulation models; understanding the impact of different rates of growth.
- *Statistics:* Understanding the importance of variability; recognizing the differences between correlation and causation, between random-

ized experiments and observational studies, between finding no effect and finding no statistically significant effect (especially with small samples), and between statistical significance and practical importance (especially with large samples).

- *Chance:* Recognizing that seemingly improbable coincidences are not uncommon; evaluating risks from available evidence; understanding the value of random samples.
- *Reasoning:* Using logical thinking; recognizing levels of rigor in methods of inference; checking hypotheses; exercising caution in making generalizations.

The differences between these topics and those found on many tests or in courses designed to meet a so-called mathematics or quantitative requirement are typical of the distinction between quantitative literacy, which stresses the use of mathematical and logical tools to solve common problems, and what we might call *mathematical literacy*, which stresses the traditional tools and vocabulary of mathematics. Indeed, it is not uncommon for a person who is familiar with a mathematical or statistical tool (e.g., the formula for standard deviation) not to recognize in a real-life situation when it should be used—or just as important, when it should not be used. Similarly, it is not uncommon for someone who knows how to use standard deviation in a specific quality control setting not to recognize the concept when it arises in a different context (such as in a course in economics).

Quantitative Literacy in Context

In contrast to mathematics, statistics, and most other school subjects, quantitative literacy is inseparable from its context. In this respect it is more like writing than like algebra, more like speaking than like history. Numeracy has no special content of its own, but inherits its content from its context.

Another contrast with mathematics, statistics, and most sciences is that numeracy grows more horizontally than vertically. Mathematics climbs the ladder of abstraction to see, from sufficient height, common patterns in seemingly different things. Abstraction is what gives mathematics its power; it is what enables methods derived in one context to be applied in others. But abstraction is not the focus of numeracy. Instead, numeracy clings to specifics, marshaling all relevant aspects of setting and context to reach conclusions.

To enable students to become numerate, teachers must encourage them to see and use mathematics in everything they do. Numeracy is driven by issues that are important to people in their lives and work, not by future needs of the few who may make professional use of mathematics or statistics. In teaching quantitative literacy, content is inseparable from pedagogy and context is inseparable from content. Fortunately, because numeracy is ubiquitous, opportunities abound to teach it throughout the curriculum. Only by encountering the elements and expressions of numeracy in real contexts that are meaningful to them will students develop the habits of mind of a numerate citizen. Like literacy, numeracy is everyone's responsibility.

Challenges of Quantitative Literacy

The penetration of numeracy into all aspects of life—from education, work, and health to citizenship and personal finance—confronts us with a rapidly evolving phenomenon that we understand at best imperfectly. Americans have had decades, even centuries, to recognize the public importance of literacy. Campaigns for literacy are commonplace, now even part of presidential politics. Yet there is little corresponding public concern about numeracy, except for ill-informed (and innumerate) obsession about SAT scores and AP calculus enrollments. The public seems not to grasp either the escalating demands for quantitative literacy or the consequences of widespread innumeracy.

Ironically, public apathy in the face of innumeracy may itself be a consequence of innumeracy. People who have never experienced the power of quantitative thinking often underestimate its importance, especially for tomorrow's society. In contrast, because it has been a staple of the school curriculum, most adults do recognize the importance of mathematics even if they themselves do not feel comfortable with it and have a highly distorted impression of its true nature. But as we have seen, numeracy is not mathematics, and public concern about mathematics education does not automatically translate into a demand for quantitative literacy.

Thus a key challenge in the campaign for quantitative literacy is to mobilize various constituencies for whom numeracy is especially important. The quality of medical care, for example, depends on numerate patients, just as wise public policy depends on numerate citizens. Educational, business, and political leaders all have a stake in a numerate public (even if they sometimes rely on the public's innumeracy to promote questionable products or policies). These leaders, however, naturally focus their attention on existing instruments such as mathematics standards, high school graduation tests, college admission tests, college placement tests, and (occasionally) college graduation requirements.

If, as seems inescapable, the importance of quantitative literacy will become ever more apparent and pressing (albeit in different ways to different groups), a second challenge is to expand these traditional instruments of educational policy to include stronger emphasis on quantitative literacy. Indeed, as the twenty-first century unfolds, quantitative literacy will come to be seen not just as a minor variation in the way we functioned in the twentieth century but as a radically transformative vantage point from which to view education, policy, and work.

THE DESIGN TEAM

This case statement was prepared by Lynn Arthur Steen of St. Olaf College on behalf of a Quantitative Literacy Design Team assembled by the National Council on Education and the Disciplines (NCED) under the leadership of Robert Orrill. Members of this team included:

Gail Burrill, director of the Mathematical Sciences Education Board at the National Research Council in Washington, D. C.

Susan Ganter, associate professor of mathematical sciences in the Department of Mathematical Sciences at Clemson University.

Daniel L. Goroff, professor of the practice of mathematics and associate director of the Derek Bok Center for Teaching and Learning at Harvard University.

Frederick P. Greenleaf, professor of mathematics in the Department of Mathematics at the Courant Institute of New York University.

W. Norton Grubb, David Gardner Professor of Higher Education Policy, Organization, Measurement, and Evaluation at the Graduate School of Education of the University of California at Berkeley.

Jerry Johnson, professor and chairman of the Mathematics Department at the University of Nevada in Reno.

Shirley M. Malcom, head of the Directorate for Education and Human Resources Programs at the American Association for the Advancement of Science in Washington, D. C.

Veronica Meeks, mathematics teacher at Western Hills High School in Fort Worth, Texas.

Judith Moran, associate professor of quantitative studies and director of the Mathematics Center at Trinity College in Hartford, Connecticut.

Arnold Packer, chair of the SCANS 2000 Center at Johns Hopkins University in Baltimore.

Janet P. Ray, professor at Seattle Central Community College in Seattle, Washington.

C. J. Shroll, executive director of the Workforce Development Initiative at the Michigan Community College Association in Lansing, Michigan.

Edward A. Silver, professor in the Department of Mathematics at the University of Michigan School of Education in Ann Arbor.

Lynn A. Steen, professor of mathematics in the Department of Mathematics at St. Olaf College in Northfield, Minnesota.

Jessica Utts, professor in the Department of Statistics at the University of California at Davis.

Dorothy Wallace, professor of mathematics in the Department of Mathematics at Dartmouth College in Hanover, New Hampshire.

Like any committee effort, this case statement represents not unanimity of views but a consensus on important issues that members of the Design Team believe are both timely and urgent.

REFERENCES

Bernstein, Peter L. *Against the Gods: The Remarkable Story of Risk.* New York, NY: John Wiley, 1996.

Buxton, Laurie. *Math Panic.* Portsmouth, NH: Heinemann, 1991.

Cockcroft, Wilfred H. *Mathematics Counts.* London: Her Majesty's Stationery Office, 1982.

Cohen, Patricia Cline. *A Calculating People: The Spread of Numeracy in Early America.* Chicago, IL: University of Chicago Press, 1982; New York, NY: Routledge, 1999.

Crosby, Alfred W. *The Measure of Reality: Quantification and Western Society, 1250–1600.* Cambridge, UK: Cambridge University Press, 1997.

Forman, Susan L. and Steen, Lynn Arthur. *Beyond Eighth Grade: Functional Mathematics for Life and Work.* Berkeley, CA: National Center for Research in Vocational Education, 1999; reprinted in *Learning Mathematics for a New*

Century, Maurice Burke (Editor), Reston, VA: National Council of Teachers of Mathematics, 2000.

International Life Skills Survey (ILSS). Policy Research Initiative. Statistics Canada, 2000.

Kirsch, Irwin S. and Jungeblut, Ann. *Literacy: Profiles of America's Young Adults.* Princeton, NJ: Educational Testing Service, 1986.

Murnane, Richard and Levy, Frank. *Teaching the New Basic Skills: Principles for Educating Children to Thrive in a Changing Economy.* New York, NY: Free Press, 1996.

National Center for Education Statistics (NCES). *Adult Literacy in America. Report of the National Adult Literacy Survey (NALS).* Washington, DC: U.S. Department of Education, 1993.

National Center for Education Statistics (NCES). *NAEP 1996 Trends in Academic Progress.* Washington, DC: U.S. Department of Education, 1997.

National Council of Teachers of Mathematics (NCTM). *Principles and Standards for School Mathematics.* Reston, VA: National Council of Teachers of Mathematics, 2000.

Organization for Economic Cooperation and Development (OECD). *Literacy, Economy, and Society: Results of the First International Adult Literacy Survey.* Paris: Organization for Economic Cooperation and Development, 1995.

Organization for Economic Cooperation and Development (OECD). *Literacy Skills for the Knowledge Society.* Washington, DC: Organization for Economic Cooperation and Development, 1998.

Paulos, John Allen. *Innumeracy: Mathematical Illiteracy and Its Consequences.* New York, NY: Vintage Books, 1988.

Paulos, John Allen. *A Mathematician Reads the Newspaper.* New York, NY: Doubleday, 1996.

Porter, Theodore M. *Trust in Numbers: The Pursuit of Objectivity in Science and Public Life.* Princeton, NJ: Princeton University Press, 1995.

Programme for International Student Assessment (PISA). Organization for Economic Cooperation and Development (OECD), 2000.

Secretary's Commission on Achieving Necessary Skills (SCANS). *What Work Requires of Schools: A SCANS Report for America 2000.* Washington, DC: U.S. Department of Labor, 1991.

Sons, Linda, et al. *Quantitative Reasoning for College Graduates: A Supplement to the Standards.* Mathematical Association of America, 1996.

Steen, Lynn Arthur. *Why Numbers Count: Quantitative Literacy for Tomorrow's America.* New York, NY: The College Board, 1997.

Steen, Lynn Arthur. "Numeracy: The New Literacy for a Data-Drenched Society." *Educational Leadership,* 57:2 (October 1998) 8–13.

Steen, Lynn Arthur. "Reading, Writing, and Numeracy." *Liberal Education,* (Summer 2000).

Tobias, Sheila. *Overcoming Math Anxiety.* New York, NY: Houghton Mifflin, 1978. Revised Edition. New York, NY: W. W. Norton, 1993.

Tufte, Edward R. *The Visual Display of Quantitative Information; Envisioning Information; Visual Explanations—Images and Quantities, Evidence and Narrative.* (3 Vols.) Cheshire, CT: Graphics Press, 1983, 1990, 1997.

White, Stephen. *The New Liberal Arts.* New York, NY: Alfred P. Sloan Foundation, 1981.

Wise, Norton M. *The Values of Precision.* Princeton, NJ: Princeton University Press, 1995.

The Emergence of Numeracy

Patricia Cline Cohen

University of California, Santa Barbara

"The Case for Quantitative Literacy" substantially advances our thinking in at least four ways. It identifies various components ("elements") of this style of thinking that together give us a comprehensive and appropriately complex definition of quantitative literacy. It then gives a multitude of examples of actions and behaviors ("expressions") occurring in daily life that call for this kind of thinking, from the simple to the esoteric. It next distinguishes the bundle of skills that constitute quantitative literacy as an academic subject. And finally, the case statement makes clear that quantitative literacy and mathematics are really two quite different things.

I wish I had possessed such a precise and nuanced statement a quarter century ago when I wrote a Ph.D. dissertation in the field of American history about something I vaguely termed a *quantitative mentality* (Cohen, 1977). My use of the term *mentality* drew on the work of a French historical school prominent in the 1960s and 1970s (led by historians Jacques LeGoff and Lucien Febvre) that championed the study of *l'histoire des mentalités,* meaning deep mental structures that persist in cultural groups over time. In contrast to the more typical historical focus on events, this kind of study explored the mental equipment, *l'outillage mental*, characteristic of a particular culture. The study of *mentalité* was sometimes thought of as the intellectual history of common people, the study of the thought patterns and fundamental attitudes of the members of a culture comprehended in the aggregate. But I found that *mentalité* was an inherently slippery concept to apply.

My interest was drawn to the subject by the realization that in early nineteenth-century America, quantitative description and numerical reasoning seemed to blossom. These were the early years of a vast transformation in the economy, and I suspected there was a connection between the market revolution, improved delivery of arithmetic education, and the propensity to use numbers to support arguments of all kinds in the realms of politics, economics, social reform, and the like. My goal was to describe and explain an important piece of mental equipment just at the moment that it was coming into prominence, but without definitional precision it was hard to draw boundaries around my study.

Numeracy in History

By the time I had moved from a dissertation to a revised book (Cohen, 1982), I had happily come upon the word *numeracy* in a British dictionary. That word, because of its parallels to literacy, helped enormously to crystallize my thinking about what constituted a quantitative mentality. The word turned my focus to the history of a skill and the specific, everyday contexts in which it was manifested, and finally clarified that I was not writing a history of mathematics. The aim of my book was to investigate selected areas in which this new comfort and familiarity with numbers supplanted previous approaches to similar problems where nonnumerical thinking had once prevailed.

There was of course no single moment in time when American society moved from being prenumerate to being fully numerate. Instead, my book traced the gradual extension of numeracy to a host of specific activities: taking censuses for military and political uses, evaluating medical outcomes using simple statistics, revamping arithmetic teaching to gear it to a new commercial order, compiling numerical facts about the state ("statisticks" as in descriptive statistics) to help statesmen govern, collecting voting statistics to improve the management of party politics in a democracy, and finally, mounting numerical arguments in the service of the reform movements of the 1820s to 1840s. Through it all, I was alert to the growing sophistication of numerical argument and to the cultural lags that held some parts of the population back from full participation in this new style of thinking.

The case statement greatly sharpens our ability to investigate the status of numeracy in modern as well as past societies. The distinction it draws between mathematics and quantitative literacy is an important one, and to the extent that our schools emphasize the former and not the latter, we fail to equip our citizenry with essential skills.

Certainly that distinction, between numeracy as a concrete skill embedded in the context of real-world figuring and mathematics as an abstract, formal subject of study, was sharply drawn in the eighteenth and early nineteenth centuries. The educated man who studied algebra, geometry, and trigonometry, and perhaps calculus in college, likely breezed through a book on the basic "rules" of arithmetic in a year's time, often learning formulaic algorithms for manipulating numbers in a first-level college course. (And I do mean *man*; rare was the woman who could advance in formal mathematics training because all of higher education was closed to women.)

Even for men, precollege work emphasized Latin and Greek, the standard admission requirements for university. Harvard University did not even require basic arithmetic for admission until 1802. Courses in higher mathematics were similarly abstract and devoid of practical application. The individual who pursued this kind of learning was ·unusual, having the money and advantage to attend college plus the talent drawing him to higher mathematics; such a person likely had high aptitude for the subject in the first place and could pick up basic arithmetic in short order.

Around 1800 the far more common exposure to arithmetic consisted of the study of practical skills aimed at boys planning to enter the mercantile life. Students on this track would spend three or four years, between the ages of 10 and 14, working through commercial arithmetic texts and practicing on real-life problems such as calculating board feet to build houses, figuring discounts and interest rates, and manipulating the vast (and highly complex and confusing) array of denominate numbers used for measuring goods. In this second track, the teaching of arithmetic was so completely framed in terms of real-life examples that a student might not fully realize that the multiplication involved in the board feet problem was the same operation as the multiplication required by the infamous Rule of Three used to figure such things as cost per yard.

Instead of learning abstract rules capable of generalization in the real world, students memorized problem after problem rooted in real-world calculations, burdening the memory—or they inscribed a copybook carried into the countinghouse life—with examples of every conceivable kind of problem. At the higher ends of this study, texts taught things like the rule of fellowship (to figure out how to divide profits or losses in partnership contracts), the rule of discount, and the rule of barter. Higher branches of mathematics with practical applications included geometry and trigonometry as applied to navigation, surveying, and gunnery. Each of these fields had separate textbooks and followed the same style of problem-based memory learning as commercial arithmetic.

Neither the formal college-level mathematics nor the rote-learned commercial arithmetic of 1800 could be described as intuitive. In fact, commercial arithmetic was so completely context-specific that it probably retarded the development of quantitative literacy. With the intensification of market activity in the United States after the War of 1812, some educational theorists proposed entirely new ways to teach arithmetic: they simplified it, greatly reduced the number of "rules" by generalizing the operations, encouraged discovery methods of learning, rearranged the order of subjects taught, and started teaching it to 5- and 6-year-olds, instead of 10- to 12-year-olds. The particular methods of teaching the "new math" of the 1820s remained controversial for many decades, but the general outcome of this new educational theorizing was clearly highly beneficial to the nation. Via the common schools and the push for universal elementary education, the new math introduced many people, males and females, to the basics of quantitative literacy.

Numeracy Today

Now, two hundred years later, we inhabit a society inundated with numbers. The number skills needed to carry on daily life activities have increased, and while we have developed workarounds to simplify some of them, for example, computers and calculators in place of the "ready reckoner" tables of 1800, the need for numerical understanding is ever greater. We particularly run a danger because of a lack of numerical sophistication in the political realm. Quantitative literacy is required to understand important political debates on issues such as Social Security funding, the differential effects of various tax-reduction plans, and health insurance options. Relatively few Americans have the quantitative savvy (and maybe also the time) to work through these policy debates and evaluate all their implications.

So we take shortcuts instead, not all of them good. Thirty-second advertisements and the quest for the perfect sound bite for the evening newscast pressure politicians to reduce their take on a complex policy to a short, clear statement, which pictures the "typical" family and its projected tax savings under candidate X or a "typical" elder citizen and her projected savings on prescription drugs under candidate Y. Lacking the quantitative literacy to make sense of policies, voters substitute evaluations of the character and vision projected by candidates, trusting that the right person will delegate policy decisions to a team of experts sharing the general political ideology of the winner.

Are any lessons to be derived from studying the history of numeracy? One clear lesson is that the methods of teaching matter greatly. Benjamin Franklin, revered for his intuitive, commonsense genius, struggled mightily in the 1730s with his commercial arithmetic lessons, only to find at a later age that he could teach it to himself with ease (no doubt drawing on his superb intuitive intelligence and his real-world experience apprenticing in a printer's business). Formal arithmetic instruction was postponed until students were 10 years old or older precisely because it was such a heavy study, as then taught. A 10-year-old would already know how to number and count before opening the text. The first wave of new math in the 1820s made it possible for children more average than Franklin (and much younger, too) to be successful in arithmetic.

Continued pedagogical improvements have enabled many students to improve their quantitative literacy. We have seen several revisions of the arithmetic curriculum in the last fifty years, and controversy now rages in states such as California over state-mandated standards. But it seems to me that the debate is still focused on the various methods of teaching number facts. The case statement moves beyond that debate in arguing that quantitative literacy matters well beyond the sphere of mathematics and science; it is indeed a basic thinking skill parallel to verbal literacy.

How then do we reposition arithmetic training to encourage a stronger emphasis on numeracy? We need to expand educational experiences conveying the message that quantitative literacy is not only about arithmetic and higher mathematics but also about a general skill (or habit of mind) that is required in many subjects across the curriculum. Courses other than mathematics need to reinforce this skill by demonstrating, indeed requiring, its use.

Along with an enlarged application of numbers must come an appreciation for what it means to approach an issue or problem from a quantitative standpoint. What is gained, and what is lost? What does the numerical argument account for, and what does it fail to include? What are the uses and misuses of quantitative thinking? My own study of early American numeracy revealed that at different times certain types of quantification were embraced while others were ignored or rejected.

Numbers are not unimpeachable facts; they can be and often are contested. Rising nation-states in the seventeenth century saw the wisdom of enumerating population (to estimate military strength), but not all of the enumerated persons agreed, in light of a strict biblical prohibition on taking censuses. Twentieth-century economists have developed the GNP as a measure of gross national product, a quantitative measure of the productive capacity of the country, but recent critics have faulted the GNP for excluding forms of productivity such as women's household labor or for ignoring the environmental costs of productivity. Every act of social science quantification has built into it a set of decisions about what to count and how to categorize. Education in quantitative literacy has to make citizens sufficiently sophisticated to be aware of such issues.

The case statement persuasively argues that education in this style of thinking is essential in the modern world. It should spark a renewed debate about the adequacy of an arithmetic curriculum that, in its broad outlines, has been in place since the 1820s.

I am especially happy to see this debate now, because when I first took on this subject very little attention was paid to the problem of numeracy either as a historical topic or as a pressing educational problem. In the early 1980s there were books on the history of the census and demographic thought in past times, but only in the last decade has there been a real flowering of work about the history of numeracy (e.g. works by Crosby, Hobart, Swetz, Hadden, Porter, Stigler, Wald, Anderson, Alonso and Starr, Desrosieres, and Poovey). Most of these efforts have focused on European thinkers and governments, but with the current spotlight on educational practices in the United States, I hope there will be renewed interest in reconstructing the history of numeracy in America.

REFERENCES

Cohen, Patricia. "A Calculating People: The Origins of a Quantitative Mentality in America." Ph.D. Thesis, University of California, Berkeley, 1977.

Cohen, Patricia Cline. *A Calculating People: The Spread of Numeracy in Early America*. Chicago, IL: University of Chicago Press, 1982; New York, NY: Routledge, 1999.

Connecting Mathematics with Reason

Joan L. Richards
Brown University

We are often told to study history to learn from past mistakes, but this is a vain hope; history does not repeat itself such that we can use it to predict the future. But avoiding mistakes does not exhaust the ways history can be used to illuminate the present. One way in which it is most useful is as a check on how we organize and understand our world. Looking at history can help us see the hidden issues and assumptions that lie behind how we talk and think about that world. The arguments that in the past have justified the place of mathematics in education may shed light on the challenges we face today as we try to understand and further quantitative literacy.

Mathematics in Eighteenth-Century France

One of the defining moments contributing to our modern view of the educational value of mathematics occurred in France at the end of the eighteenth century. In the immediate post-revolutionary period, the education of the new French citizen was a major preoccupation. From the Enlightenment came the conviction that man was a rational animal; teaching mathematics was of major importance in the post-revolutionary educational program to strengthen the reason of the new *citoyen,* or citizen. A form of quantitative literacy that linked mathematical instruction with reason made mathematics an integral part of the curriculum in the *Écoles Centrales,* the short-lived revolutionary schools of the 1790s.

The rhetoric surrounding these schools was stirring, but its institutional manifestation was short-lived; by the time Napoleon came to power, the

Écoles Centrales were defunct. Napoleon did not abandon education, however. The system of *Lycées* that grew up under his administration persists to the present day. But the *Lycées* were more classically oriented than the *Écoles Centrales;* mathematics did not hold pride of place in their curriculum.

This does not mean that mathematics was not pursued in early nineteenth-century France, however. The subject was of central importance at the elite *École Polytechnique,* where Napoleon's engineers were educated. Mathematical tests decided admission to the school and determined rank within it. No longer an essential part of everyone's education, in the first decades of the nineteenth century knowledge of mathematics had become a mark of status and social differentiation.

There were real benefits to this elite status. In the first decades of the nineteenth century an extraordinary group emerged at the *École Polytechnique,* free to pursue mathematics without having to justify or explain their work to a large public. The first quarter of the new century saw Jean Victor Poncelet develop projective geometry, Pierre Simon LaPlace systematize probability theory, and Jean Fourier develop Fourier series.

Mathematics' change in status, from universal reason to intellectual sieve, did more than encourage research, however; it left its mark on the subject itself. This can be seen in the work of yet another *École* mathematician, Augustin Cauchy, who in his *Cours d'Analyse* of 1822 firmly established calculus on the rigorous basis of the limit. It was a signal achievement to bring rigor to a subject that had been poorly understood throughout the eighteenth century, and his work has long been hailed as a classic. Less well known, though, is that Cauchy's students rioted violently in protest against his work and his teaching. From their point of view, Cauchy's rigor was an assault on the humane mathematics that had been touted by the revolutionaries of the 1790s. The students argued that although Cauchy brought rigor to calculus, he did so at the cost of reasonableness. Another way to put it is that he introduced a course in rigorous mathematics at the cost of the kind of quantitative literacy that had been advocated and taught by an earlier generation of mathematicians.

In this particular case, the students may be said to have won the battle— Cauchy never published his projected second volume—but they certainly lost the war. Throughout the nineteenth century in France, as in Germany, rigor rather than reason was the preeminent value in mathematics. The case for mathematics as the route to reason was more solidly built in England, however, where throughout the century mathematics played a central role in the development of the idea of a liberal education.

Mathematics in Nineteenth-Century England

Much of the English discussion about mathematics and education was focused at Cambridge University where, until the 1860s, everyone who wanted a degree had to pass a major mathematics examination. Over the course of the century this examination, known as the Mathematical Tripos, became ever more grueling and competitive. This meant that during the 1830s, 1840s, and 1850s, much of England's intelligentsia spent their time in college pursuing evermore sophisticated mathematics; those who placed high in the Tripos were certainly the mathematical equals of those graduating from the *École Polytechnique*.

Beyond technical proficiency, however, comparison is difficult because mathematics was not the same subject in France and England. Ultimately the education at the *École Polytechnique* was more about training engineers than about educating either mathematicians or citizens. Cambridge University, for its part, was an institution whose major mission was to educate the clergy of the Anglican Church. Mathematics was pursued there as a form of quantitative literacy, a way to teach young men to understand and recognize the truth.

Cauchy's form of rigor was not clearly compatible with this goal; to this day there is no English translation of the *Cours d'Analyse*. This does not mean that the English were ignorant of his work, however; within two decades many had adopted his limit-based approach to calculus. The difference in the English and French understanding of that approach can be seen by contrasting Augustus De Morgan's 27-page chapter explaining the concept of the limit to Cauchy's two-line, operational definition. It can be further illustrated by the two men's approaches to divergent series. Because these often led to ambiguous results, Cauchy ruled them out of legitimate mathematics. De Morgan, however, was quite comfortable using such series: "Divergent series [were] nearly universally adopted for more than a century," he noted in 1844, "and it was only here and there that a difficulty occurred in using them" (Richards, 1987). In De Morgan's view, the creative possibilities of mathematics lay precisely in the challenges posed by this kind of ambiguity; to define them out of the subject because of a concern with rigor was nothing but counterproductive.

Over the course of the nineteenth century, the view of mathematics as reason remained strong in England, but it was challenged on at least two fronts. On the one hand, there was an internal challenge. It arose at Oxford University, which was, like Cambridge, devoted to educating the clergy. At Oxford, however, the curricular focus was on logic rather than mathematics. In the 1830s, the leading minds at these two universities faced off in a major battle over whether logic or mathematics was a better way to teach young men to use reason. The issue soon came down to whether the goals of education were better achieved by teaching the rules of reason, as in logic, or by practicing reason, as in mathematics— whether it was better to learn reason by precept or by practice.

When the battle was originally joined, neither side questioned whether mathematical arguments were quintessentially reasonable, or whether the forms of logic described that reason. Over the course of the next several decades, however, some began to ask whether even the time-honored mathematical proofs of Euclid were wholly satisfactory and, even if they were, whether Aristotle's logic was adequate to the reasoning they embodied. The easy connections among mathematics, logic, and reason on which rested the neo-Enlightenment program of mathematical and logical teaching became ever less clear over the course of the nineteenth century.

At the same time the English were questioning the relationships among mathematics, logic, and reason, they were adapting to a more external challenge in the form of professionalism. By the second half of the nineteenth century, the elite, self-defined communities of mathematicians found on the continent began to pose a powerful alternative to the gentlemanly ideals on which the English liberal education rested. By the end of the century, mathematics was beginning to be recognized as a specialized research subject in England. Thus, by early in the twentieth century, the problems inherent in identifying mathematics with reason were leading in England, as they already had on the continent, to the pursuit of "pure" mathematics and "formal" logic. By isolating their subject in this way, mathematical practitioners freed themselves from the confusions encountered at the interface of mathematics and reason.

This meant that practitioners could pursue their research in peace. But, at the same time, it strained their claims to model reason—whether by precept or by practice. Formal logic may be pure, but many were not willing to concede that it describes how we actually think; abstract mathematics may be rigorous, but many argued that it does not model human reason. As mathematics and logic were being redefined and purified in one context, they became increasingly irrelevant in another. By the beginning of the twentieth century, mathematics in England had become a research subject, but it no longer held pride of place in the Cambridge liberal arts curriculum.

The issues that face contemporary educators as they try to define and clarify the place of mathematics in the U.S. educational system are reminiscent of the ones that faced those defining the curricula at the *Écoles Centrales,* or at nineteenth-century Cambridge. But they are not the same. If a single lesson is to be drawn from the interrelated stories of mathematics and education in nineteenth-century England and France, it is that the link between mathematics and reason has always been a dynamic one, that both sides of the equation are highly susceptible to the vagaries of time and place. Both Cauchy's students and his English contemporaries recognized that mathematics as he defined it was not the same as that pursued by his mathematician forebears. Ideas of reason, as embodied in educational institutions, have changed as well. Although we, like our English predecessors, are focused on educating productive people, we are more likely to describe such people as functioning citizens or savvy consumers than as clear-thinking, moral gentlemen. What ties us to our past is the conviction that there is a powerful connection between mathematics and reason.

Quantitative Literacy

The concept of quantitative literacy is rooted in the connection between mathematics and reason. As described in "The Case for Quantitative Literacy," quantitative literacy is tied to two rather different concepts—numeracy and reason. With respect to the first, teaching quantitative literacy could perhaps be addressed relatively specifically and in a piecemeal fashion. With respect to the second, however, the current call for quantitative literacy harks back to the Enlightenment call for reason, echoes the challenges that faced the *Écoles Centrales* and nineteenth-century Cambridge, and radically expands the implications of teaching numeracy.

When teaching mathematics is seen as a way of teaching people how to think, it can no longer be isolated. Its implications spread throughout the curriculum and it has a place in every class.

In my view, that is where mathematics belongs: at the very heart of the educational project. But in advocating this position, I recognize that it places a significant responsibility on mathematicians and mathematics educators. De Morgan's defense of divergent series did not earn him respect within the mathematical community, but he stood his ground because, for him, mathematics models reason, and the ambiguities of divergent series were essential to his view of the way people think. We face similar choices as we consider what to include in a mathematics that models reason. We not only must reconsider the role of rigor but also determine the degree to which we are willing to allow machines to model reasoning for us. The historical record does not give answers to such questions, but it does place us in good company as we struggle to deal with them.

REFERENCES

Richards, Joan L. "Augustus De Morgan, the History of Mathematics, and the Foundations of Algebra." *Isis* 78 (1987), 184.

Numeracy, Mathematics, and General Education

An Interview with Peter T. Ewell

National Center for Higher Education Management Systems

"The Case for Quantitative Literacy" argues that quantitative literacy (QL) is not merely a euphemism for mathematics but is something significantly different—less formal and more intuitive, less abstract and more contextual, less symbolic and more concrete. Is this a legitimate and helpful distinction?

I believe that the distinction between quantitative literacy (QL) and mathematics is indeed a meaningful and powerful one. For me, the key area of distinction is signaled by the term *literacy* itself, which implies an integrated ability to function seamlessly within a given community of practice. Literacy as generally understood in the verbal world thus means something qualitatively different from the kinds of skills acquired in formal English courses. For one thing, it is profoundly social, and is therefore a moving target because its contents depend on a particular social context. For instance, it is easy to imagine literacies being quite different from one another in different historical periods or cultural contexts. So a literacy is not just an applied version of a discipline. Instead, it would seem to flow out of a specific set of symbolic and communication needs embedded deeply in a particular social environment or community of practice.

Another important point is that literacies are for the most part practiced invisibly and subconsciously by members of a community, not pulled out selectively and applied deliberately to a particular set of circumstances. In practicing QL, therefore, we would expect that an individual would not consciously say "Oh, this is mathematics" and enter a different ("learned") way of thinking and acting. Instead, he or she would simply act competently without invoking a disciplinary context at all. A final and related point is that, although it may have different aspects such as prose, document, and quantitative, "literacy" is really all one concept. Thus QL is presumably practiced together with other literacies in most actual circumstances, whereas mathematics as a discipline can be practiced on its own.

For all these reasons, I think that the case can indeed be made that QL is different from mathematics as customarily understood. But I also believe that these differences are not easy for most people to grasp at first. This may be in part because they are not always able to recognize their own use of quantitative concepts and tools in everyday life. It also may be in part because early exposure to mathematics presents it as a distinctly different activity from natural forms of communication. Reading and writing thus appear to be expected extensions of everyday life in ways that are not necessarily true of mathematical concepts.

For example, the notion of approximation is inherently legitimate in verbal expression: we choose words to get a point across, and the particular form in which this occurs—unless it is grossly inappropriate—rarely inhibits people from communicating with one another. In contrast, I suspect that most people's early exposure to mathematics strongly imprints the idea that it is somehow illegitimate to improvise and approximate in the quantitative realm: things are either "right" or "wrong" and must be "precise" to be of any use. As a result, I suspect that the ready analogy of quantitative facility to other forms of literacy is not apparent to a lot of people.

The idea that everything in mathematics is either right or wrong causes many students great difficulty because they prefer to think in shades of gray, not just black or white. This helps explain why mathematics serves as a "critical filter" that blocks students with weak mathematical skills from rewarding careers. Just how important is it that all students master formal mathematics? Might context-rich quantitative literacy be a more reasonable alternative?

Certainly I grant the premise of this question—that mathematics coursework does in many cases largely determine student entry into particular disciplines and careers. But I'm not entirely sure that QL in all cases represents a substitutable alternative to formal mathematical training. We're really talking about different things here, and both may be important for certain college majors and careers. All students, regardless of major or career aspiration, need context-rich QL as an integral part of their education. But other mathematical topics and contexts not explicitly addressed by the case statement may also be required for success in particular fields.

The problem here, I believe, lies much less in the distinction between QL and formal mathematics than in the fact that particular topic areas within the latter may require a different kind of treatment. Students in courses of study that generally require calculus as a prerequisite, engineering and physics for example, probably do not need to master everything that is typically addressed in a traditionally taught calculus course. But they do need to gain formal and operational mastery of particular concepts and tools that would not properly be considered part of the general QL domain. Similarly, students entering business and social science programs (as well as biology and medicine) ought to know far more about probability and statistics than is addressed by QL, at least as described in the case statement, but they may not need to be familiar with all the topics generally covered in a college statistics course.

Thus the pipeline problem for me is more about the way regular college mathematics courses typically are organized and taught than about the more basic distinction between QL and formal mathematical training. At minimum, addressing the question as posed requires (1) more modularization to allow tailored prerequisite experiences to be offered (perhaps "just in time" as students encounter particular discipline-related applications) and (2) far more experience with applications and real-world problem solving than is generally provided. As I understand it, "reform calculus" points in this direction, but I don't see these kinds of applications much in courses such as statistics or college algebra. Certainly, an enhanced and universally required QL component might prepare students to do well in such redesigned college mathematics courses, which would in turn provide more effective preparation for later disciplinary work.

At the same time, greater emphasis on direct application and understanding—regardless of the concepts being taught—would go a long way toward alleviating classic "math anxiety" for many students. The notion of QL is especially helpful here because it emphasizes that quantitative concepts are a part of everyday life. This linkage might build confidence for many students, because they could be shown to already possess some understanding, no matter what level of QL they currently have attained. Stressing the connections between quantitative tools and familiar situations would similarly reduce the feeling that numbers are part of an impenetrable foreign language.

In short, I believe most QL skills should be fully developed as a prerequisite to postsecondary education, though, like appropriate verbal and writing skills, this cannot be assumed in the short term. Students entering college with particular major and career aspirations, however, need solid backgrounds in selected areas of formal mathematics as well as QL, both taught in ways that make them applicable and engaging.

Standards and Assessments

Many people share your desire that QL skills be acquired by the end of secondary school. Do you think that current efforts to improve public education by developing state-based standards and assessments are likely to lead to improved QL skills?

I believe that depends a lot on the standards and assessments. In general, I like the direction taken by such efforts as the New Standards project, in which the statements put forward are explicit and performance-based, enabling both teachers and the public to gain some idea of what they mean. But to remain true to quantitative literacy as described in the case statement, standards really must be described in "ability" terms rather than in content or knowledge-based terms. The approach also needs to be one in which each standard is clearly illustrated by reference to the types of real-life, concrete problems students are expected to formulate and address.

This, in turn, means that the assessments used in such an approach must be authentic and complex—for instance, requiring students to shift contexts in applying concepts and to operate in real-world settings. More profoundly, in contrast with what (I think) is prevalent in K–12 practice, the emphasis should not simply be on passing high-stakes exit tests, however well constructed. Instead, standards of the type I am thinking of must be embedded in routine faculty assessments of students' classroom work, and students should be fully aware of what constitutes "good performance" on such tasks.

We can all hope that the standards movement will, somehow, lead to improving the QL skills of secondary school graduates and college entrants. But this raises a related question: Just how important is quantitative literacy in the priorities of colleges and universities? Should QL be required as prerequisite to admission to a four-year degree program? Is there any consensus on the level of QL that a college should require as part of general education?

I believe QL is an extremely important topic for colleges and universities to address. My basic position is that a high level of QL should be a condition of college admission. As I previously admitted, however, this is unrealistic given current levels of mastery—an unfortunate situation that is just as true of more established forms of literacy.

I do not believe, however, that most college faculty share the same degree of anxiety with regard to shortfalls in QL that they have with respect to verbal skills, especially writing. This is a particular problem because it tends to reinforce the kind of disciplinary "steering" already noted in your questions. To put it bluntly, many college faculty either do not recognize current shortfalls in the QL skills of incoming students or do not consider such skills to be important for later success in many fields of study. In my experience, faculty in the sciences and in some of the social sciences tend to complain about deficiencies in both quantitative and communications skills among incoming students, while those in the humanities and many other social sciences are concerned only about the latter. Even fewer faculty would embrace the idea that a particular level of quantitative facility is an integral part of aesthetics or civic life, as emphasized in the case statement. This is the biggest problem with respect to achieving consensus that some level of QL ought to be a requirement for college admission or a required component of general education.

That said, I don't believe that it would be difficult for interested college faculty to quickly arrive at consensus about domain content and an appropriate level of QL as part of a general education requirement. Indeed, I think the case statement does an admirable job of identifying the particular topical ingredients that such a requirement might contain. Given the typical distributed politics of general education at most colleges and universities, those uninterested would likely let those interested settle the "math part" among themselves as long as the resulting requirement remained roughly the size and scope of its predecessor.

I also don't believe that this is the only way to structure general education. Indeed, the past fifteen years or so has seen the emergence of many interesting alternative designs for general education that might allow more systematic attention to QL. One emphasizes problem-based courses that embed students immediately in practical settings or tasks, the nature of which automatically raises issues of QL together with other cross-cutting literacies and abilities. MIT's course on "Time," which systematically examines differing concepts of time and culminates with students building a workable clock, is a classic example. Other designs essentially turn the standard curriculum upside down by offering major-type courses early, and teaching literacy skills such as QL on an "as-needed" basis as particular applications and contexts come up. Still others define core abilities such as those contained in QL from the outset and interlace them throughout the curriculum—a structure best exemplified by Alverno College. These examples give me confidence that we are capable of constructing curricular designs for general education that can address the problem of coherence, but advocates of QL will need to recognize that they must make common cause with the proponents of other important literacies to make the case for such designs.

Why do you think college faculty outside the natural and social sciences are reluctant to support quantitative literacy? Does their reluctance represent a realistic assessment of the future needs of students or the legacy of their own educational experiences?

I don't think faculty outside the sciences and social sciences are reluctant to support quantitative literacy as much as they are indifferent to it for one reason or another. As my earlier answer suggested, most faculties tend to see both communication and quantitative literacy in "prerequisite skills" as opposed to "educated person" terms. That is, they view such instruction (typically provided by freshman-level courses in English or mathematics) largely from the perspective of what skills they, as faculty teaching subsequent courses, desire their students to have.

Humanities faculty are therefore very disturbed by deficiencies that will prevent student progress in their own courses and, at least at the moment, these are unlikely to include QL skills. Often, in fact, they see these deficiencies as things they will have to remediate themselves, at the expense of material they planned to teach. Social sciences and natural sciences faculty, on the other hand, view quantitative deficiencies (at least in part) from the same perspective that English faculty view deficiencies in writing—as something that they will have to do something about to get on with their own business. I'm not sure that either camp really views any of this from a "societal needs" perspective.

The problem is in many ways just as deep for other literacies. Writing faculty, for instance, often have a hard time understanding that the developmental paradigm of multiple cycles of "write and revise" that they teach does not always correspond well to a real world in which initial drafts must be quick and to the point. Similarly, both hard scientists and historians often rebel at "topics" courses that tend to treat their subjects not as rigorous courses in their own right but as tools for the functional citizenship or cultural literacy needed to understand contextually rooted, real-world issues. In fact, what they take to be this "lite" view of their own disciplines is often what they think they have a duty to root out. So the notion of embedding QL in general education is part of a larger issue of what role this portion of the curriculum ought to play in the first place. As I noted earlier, alternative conceptions of general education, conceptions based on interdisciplinary or problem-based courses and consciously structured to promote and reinforce key abilities such as QL as part of multiyear sequences of learning experiences, hold considerable promise for alleviating this situation, but they are far from the norm.

A Broad View of Quantitative Literacy

Setting campus politics aside, I'm curious what you think about the actual importance of numeracy. How important is quantitative literacy for understanding the arts, humanities, and public affairs? Are there any significant differences between what tends to be taught in mathematics courses and what you would expect of a numerate citizen?

I believe that QL as described in the case statement is integral to a deep understanding of all academic fields and, indeed, constitutes a condition for real intellectual discourse. But missing for most of the students (and instructors) that I encounter at the college level are three basic faculties that the statement addresses: (1) the ability to "see" mathematical functions and quantitative relationships in a graphic or structural form (and the reverse), (2) a reasonable sense of probability (manifest at a minimum in the ability to distinguish reasonable propositions from typical bookmakers' odds), and (3) the ability to estimate or approximate an answer to a multistep problem involving one or more shifts in order of magnitude. I'm not as familiar as I ought to be with what is typically taught in freshman-level mathematics courses, but I don't think these abilities feature prominently.

With regard to broader support for QL, I think conditions differ substantially among the various disciplinary families noted in the question. With respect to public affairs (and here I admit to being trained originally in survey research and econometrics), I think there is a lot of support for the position that facility in interpreting graphic representations of data, understanding basic notions of statistical confidence, and being able quickly to recognize inappropriate uses of data to support a public policy position constitute critical aspects of functional citizenship. For the arts and humanities, though, I think that the case is more difficult to make for both college faculty and the general public. The case statement makes a reasonable attempt to provide points of connection in the arts. At the same time, an evolving quantitative sense is part of the story of history and technology, which students will likely understand.

For me personally, the most compelling rationale for serious attention to QL is in some ways revealed by your question's (almost unconscious) reference to "significant differences." As I take the meaning of the question, your use of the term is analogous, rather than strictly mathematical. As such, it constitutes an excellent illustration of a broader view of QL itself. In just the same way, I recall being asked in a senior honors oral many years ago to connect the concept of the derivative, which had just come up in an interrogation about what I had learned in the required calculus and analytic geometry sequence, with my fascination with rapid patterns of societal change in Germany at the turn of the century, which had been presented in a course in which I really thought I had "learned" something. The resulting "aha" in my head at that time remains one of the most powerful connection-making experiences of my college career. This analogic use of quantitative concepts is, quite simply, not currently taught in formal mathematics courses—or, indeed, anywhere at the college or secondary level. Emphasizing it more consciously in college course work could, I believe, powerfully deepen both historical and aesthetic understanding.

> *Although the use of mathematical language as analogy and metaphor in ordinary discourse is widespread, it is rarely if ever addressed directly in formal courses. In fact, mathematical scientists tend to be very critical of what they see as "sloppy" uses of precisely defined concepts. Does this difference reflect another aspect of the "two cultures" divide? What do you think can be done to bridge it?*

In telling the story about my senior honors exam, I took a risk of precisely the kind you speak. I thought a lot before putting it before a mathematical scientist because of that "typical" reaction, but I think the root of the answer to the question you pose is that such risks have to be taken to bridge the gap, which I do think is there.

We in the nonmathematical world often are very reluctant to take such risks because of the "one-down" attitude that we feel about being "imprecise" whenever something faintly quantitative comes up in the presence of a mathematical scientist. It is safer just to avoid the subject. (Let me hasten to add that, at least in my experience, this attitude is not projected by the mathematicians in any active way; it comes instead from the perceived aura of the discipline felt by those of us outside it.) Certainly, all of us experience a version of the same concern you describe with respect to our own disciplines when terms and concepts are taken out of context and applied to things with which they have no business at all.

But surely the objective is less to root out all analogical uses of disciplinary terms than to teach students to distinguish good analogies from bad ones and, perhaps more tellingly, to be able to recognize and articulate precisely the places where a particular analogy works and where it fails. From a teaching perspective, moreover, it seems to be exactly such situations that define "teachable moments" in which an imprecise use can be probed to see if the underlying understanding is really there or, if it isn't, to try to develop it.

You raise an interesting point by noting that such uses of concepts and terms are "rarely addressed in formal courses." I'm not sure they should be—or even could be. Instead, I think they almost always arise in cross-disciplinary discourse or in practice settings in which folks are trying to get a handle (and almost any handle, at first) on a complicated, ill-defined problem. Rather than trying to engineer topics like this that you can't really teach in any formal way, we need to shape courses in many fields to make sure that such situations arise frequently, and then see where they lead.

Similarly, as faculty, we need to take risks in stretching concepts out of context and testing the resulting uses on one another when trying to stake out common ground in general education. Otherwise, by the logic of curricular politics I outlined above, we end up entirely isolated from one another. Taking risks in the presence of students—for instance in introductory interdisciplinary or problem-based college courses—is a powerful introduction to what academic discourse ought to be about.

Quantitative Literacy Across the Curriculum

Speaking of faculty taking risks, is it reasonable to teach quantitative literacy "across the curriculum" as writing often is taught, or does it require special expertise? Which teachers would be best suited to help students become quantitatively literate?

Yes, it does make sense to teach QL across the curriculum. Indeed, I can't conceive of any other way it could be done effectively without turning it into a "discipline" instead of a "literacy," but the analogy with writing also points out some of the substantial difficulties involved, especially in a college setting.

One is the notion of "special expertise" mentioned in the question. A major challenge in implementing writing across the curriculum, for instance, is the fact that faculty do not all know automatically how to coach or assess writing effectively, so substantial efforts at faculty development are generally required. I believe that the same level of effort is required for quantitative literacy and needs to be dedicated to both mathematical scientists and faculty in other disciplines.

The parallel with writing also suggests the need to consciously structure assignments and exercises across the disciplines so they simultaneously reflect meaningful specific subject-area applications and reinforce agreed-upon cross-disciplinary QL skills. This implies a "matrix" design for the curriculum in the early college years—something quite compatible with efforts such as freshman learning communities or similar linked course approaches.

Regarding who should teach QL, I see need for two kinds of expertise. One, of course, is provided by faculty in the quantitative disciplines, especially in areas such as business, science, and the social sciences. Many of these faculty already demonstrate QL in their research and writing. Further development efforts might allow them to more consciously model it in their classrooms and embed it in student assignments. Another type of expertise is probably also required, however, in the form of more specialized "QL coaches" to staff mathematics labs and instructional development centers. These individuals, like writing coaches, would have to be specially trained to assist faculty in developing the kinds of pedagogy and materials best suited to foster QL and to help students overcome typical difficulties with QL. This second role probably requires a unique kind of preparation and individuals with a typical mathematics background might or might not be best suited for it. The trick here, as in the case of writing, is to build an attractive career path for such individuals because they will be largely outside the disciplinary mainstream.

Most of what we've talked about in this interview concerns QL as a pedagogical or curricular issue associated with teaching college students. I wonder, in conclusion, if you have any thoughts about QL as a broader issue of public policy? Should it be such an issue? And, if so, how might it be raised and sustained?

I'm glad you asked that because our work at the National Center for Higher Education Management Systems (NCHEMS) often involves discussions with public policymakers and business leaders about the skill sets that ought to characterize the workforce and citizenry of the twenty-first century. And I'm happy to say that most of the folks we interview on these topics see a strong role for the kinds of abilities discussed in the case statement. As you might expect, we see this most prominently in the employment community, for two reasons. First, more and more workers, even at entry level, are encountering technology and thus need the ability to comprehend manuals and training materials that require strong numeracy and quantitative skills. Second, more and more businesses are using quality management processes that require every worker to acquire a basic understanding of such concepts as sampling, variation, and significant difference. Together these forces mean strong advocacy for QL skills as part of a larger literacy package.

What I find particularly interesting about the NCHEMS discussions, though, is that they strongly reinforce the point I made at the outset: as a "literacy," QL is not practiced in isolation nor can it be separated from a particular social context. Indeed, we find that most employers and policymakers have a hard time making lists of the attributes they want to see in workers and citizens without subtly combining things that academics of all types like to separate, such as "verbal and quantitative," "cognitive and attitudinal," "academic and vocational." It's a refreshing perspective that, at least to me, constitutes one of the strongest validations of the concept of QL itself.

Reflections on an Impoverished Education

Alan H. Schoenfeld

University of California, Berkeley

> *"When I use a word," Humpty Dumpty said, in a rather scornful tone,*
> *"it means just what I choose it to mean—neither more nor less."*
>
> *"The question is," said Alice, "whether you* can *make words mean so*
> *many things."*
>
> *"The question is," said Humpty Dumpty, "which is to be master—that's*
> *all."*
>
> *(Carroll, 1960, 269)*

If ever there was one, "quantitative literacy" is a case in point for Alice and Humpty Dumpty. As "The Case for Quantitative Literacy" makes abundantly clear, that phrase means many things to many people. Indeed, as often happens with complex ideas, the phrase may mean different things to the same person at different times. There I serve as a prime example.

School Mathematics

I grew up in the days when "mathematics" meant "school mathematics," divorced from the real world. My high school algebra course consisted of pure symbolic manipulation, for example, and my geometry course focused on producing proofs. Similarly, although my advanced algebra, trigonometry, and precalculus courses contained a few problems that pretended to deal with real-world phenomena, for the most part the contexts of these problems were so idealized as to render any applications meaningless. (When I was a student, probability and statistics, likely locuses of real-world problems, were nowhere to be seen in the secondary school curriculum.)

There was also extensive tracking. The main track aimed for college mathematics. Those not on the college preparatory track were shunted to dead-end courses in shop math or business math that offered neither "real" (that is, abstract, college preparatory) mathematics nor useful skills. Students who left the college preparatory track typically left mathematics as soon as they could. From ninth grade on the attrition rate was roughly fifty percent: only about half the students who completed mathematics at any grade level enrolled in a mathematics course the following year.

The same held in college, only more so. My undergraduate courses in probability and statistics dealt largely with probability distributions; the real world was not really present. Later on, in the 1970s, when as a new faculty member I taught the first generation of "new, improved" courses in elementary statistics, they too were removed from what most people would consider everyday reality. Hypothesis testing, for example, typically dealt only with narrow questions such as whether a batch of ball bearings produced at a factory was defective. To make things worse, the relevant numbers (means and standard deviations for sample sets of ball bearings) were given to students. The only computational tools available were paper and pencil, so working with real data sets was out of the question.

In sum, the only mathematics studied in my day was abstract, formal school mathematics. The real world (and thus quantitative literacy) was something else altogether. Of course, I lived in the real world. There I regularly used mathematical ideas, although not in any way that was obviously derivable from my formal training. I designed bookcases that had to fit exactly right. I made major purchases and worried about the accumulation of interest. I had to make sense of quantitatively based claims in the media. For example, was the chemical alar, which had been used on apple crops, really dangerous? Did electric power lines cause radiation damage? And what about major political issues? The stakes were large, the costs of quantitative illiteracy enormous.

Consider, for example, Ronald Reagan's "voodoo economics." Challenged in the 1980 presidential debates to explain his budgetary proposals, Reagan responded by waving his arms: "There's a line that goes like this [moving his arm in an upward direction, from left to right] and another line that goes like this [moving his arm in a downward direction, from left to right]. When those two lines cross, we'll have a balanced budget." This was unmitigated nonsense. Who knows what the lines represented, what the point of intersection meant? It did not matter. Perhaps intimidated by the mathematics, or charmed by his performance, interviewers did not follow up. The next day, newspapers reported that Reagan won the debate on economics.

The consequence, eight years later, was that the United States had gone from a budgetary surplus to the largest national debt in history. Some years later, California voters opted to invest in prisons rather than schools; Californians will pay the price for that decision in the years to come. Although it may be stretching the notion of quantitative literacy a bit, the fact is that a trends analysis would have pointed out the difficulties with this kind of policy. The absence of mathematical sense-making makes a big difference in the real world. (For a series of vignettes describing the disjunction between the ability to do formal reasoning, as in school mathematics, and the ability to use mathematics as a form of sense-making, see Schoenfeld, 1990.)

Quantitative Literacy

So quantitative literacy counts—big time. Generally speaking, I am comfortable with the description in the case statement. It is clear to me that quantitative literacy includes the various elements described in the statement: confidence with mathematics; a cultural appreciation of mathematics; the ability to interpret data, to think logically, to make decisions thoughtfully, to make use of mathematics in context; and more. Likewise, the expressions and skills seem well chosen. The case statement is entirely consistent with my general sense of what quantitative literacy should be: the predilection and ability to make use of various modes of mathematical thought and knowledge to make sense of situations we encounter as we make our way through the world.

Of course, that definition still begs a host of questions. Does quantitative literacy differ from what we learn, or should learn, in mathematics classes? Should we test for it as an exit skill from high school? From college? Can we identify certain courses that might serve as proxies—if you pass the course, then you meet the quantitative literacy requirement? In the context of my previous experience, classroom realities, and demographic data, these all at one time seemed to be very reasonable questions.

Convergence

I have now come to think about these issues differently. Recently I served as one of the writers of the National Council of Teachers of Mathematics' *Principles and Standards for School Mathematics* (NCTM, 2000). That experience—the goal of which was to outline a vision of mathematics education for the decade to come—provided me with the opportunity to reconceptualize my views of mathematics instruction and, concomitantly, of issues related to quantitative literacy. I now believe the following:

1. In the past, "quantitative literacy" and "what you learn in mathematics classes" were seen as largely disjoint. Now, however, they should be thought of as largely overlapping.

2. Every student should be enrolled in mathematics courses every year he or she is enrolled in high school.

3. Over the four years of high school mathematics, all students can and should become quantitatively literate and learn the mathematics that will prepare them for college.

I believe these three goals are both reasonable and desirable. Let me explain why. The reasons have to do with a convergence of the needs of the general citizenry for quantitative literacy and the needs of those who will ultimately pursue careers in mathematics and the sciences.

Once again, I will use my own experience as a case study. My Ph.D. is in mathematics; by most standards, I was very "well trained." Nonetheless, the mathematics education that I received was in many ways impoverished. Let me count the ways.

First, it was not until I was long into my career as an undergraduate that I encountered any situations that could really be called "problem solving." Most of the tasks I was assigned consisted of the application of tools and techniques I had just been shown. The idea of confronting a situation and making sense of it was not part of my education. Nor was learning any of a wide range of problem-solving techniques (e.g., heuristics).

Second, the mathematics I studied was "pure"; nary an application was to be seen. I never confronted an ill-defined situation, decided which aspects of it were inherently important, characterized them mathematically in a model, analyzed the properties of the model, drew conclusions about the situation on the basis of the model, or analyzed the reasonableness of those conclusions. I never had the opportunity to critique such models, or discover what makes for a good model and what makes for a bad one.

Third, I neither saw nor worked with any data in high school, and I worked only with "cooked" data in college. Fourth, with the exception of producing formal proofs on demand, the extent of the mathematical communication that was required of me was to produce a series of scribbles and to draw a box around the correct answer. I would have been a much better-trained mathematician if each of these issues, and others, had been addressed.

Interestingly, every one of the items I would have profited from as a mathematician-in-training is absolutely essential for literate citizenship. First, everyone needs to be an adaptive learner and problem solver. In the real world, problems do not come neatly packaged with methods of solution attached; our job is to figure out how to approach them. Second, as the case statement makes clear and as I argued above, literate citizenship calls for making a plethora of informed decisions—about interest rates, about situations that are inherently probabilistic, about the nonsense spewed by politicians. The best way to learn to make sense of applied situations, and to learn to assess claims made by others, is to have lots of practice building and assessing mathematical models.

Third, not only are we inundated by data but we now have access to technologies and techniques that enable us to operate on real data sets. Students can gather and analyze sets of data from real-world situations. Such skills will prepare them for grappling with data when they need to and for interpreting data that confront them. Fourth, it has long been understood that getting the right answer is only the beginning rather than the end of being effective on the job. The ability to communicate our thinking convincingly is equally important. Where better than in mathematics classes to learn this skill?

In short, the mathematical skills that will enhance the preparation of those who aspire to careers in mathematics are the very same skills that will help people become informed and flexible citizens, workers, and consumers. Moreover, a fair amount of mathematics can be motivated by interesting problems and learned in the process of solving them. Our goal should be to build a solid core of mathematics instruction that will serve both the mathematical and quantitative literacy needs of all students, while providing a solid base for those who desire the further study of mathematics.

Who should be responsible for this instruction? In the best of all possible worlds, instruction in all subject matters should touch on quantitative literacy whenever appropriate. Patterns of sense-making in science are often heavily quantitative, likewise in the social sciences. The study of history has been transformed through economic (i.e., quantitative) analyses. Questions of authorship—for example, did Shakespeare really write a particular play?—have been addressed by examining whether the frequency of word usage in the play in question differs significantly from Shakespeare's word usage in his well-known plays. It would be lovely to see "quantitative literacy across the curriculum" join "writing across the curriculum" as an accepted responsibility of all those who teach, at all levels.

My guess, however, is that writing across the curriculum is typically more rhetoric than reality and that infusing quantitative literacy throughout the curriculum would be an uphill battle, viewed by some as cultural imperialism. Thus, while inviting help from colleagues in other disciplines, I see the *Principles and Standards for School Mathematics* as the primary mechanism for achieving broad-based quantitative literacy. Four years of appropriately designed mathematics should ensure that all students emerge from high school quantitatively literate and prepared to pursue further study of mathematics if they so desire. For students who go on to postsecondary education, additional doses of mathematical sense-making—whether encountered in college mathematics or in other courses—couldn't hurt.

REFERENCES

Carroll, Lewis. *The Annotated Alice (Alice's Adventures in Wonderland & Through the Looking Glass)*. Introduction and notes by Martin Gardner. New York, NY: Bramhall House, 1960.

NCTM. *Principles and Standards for School Mathematics*. Reston, VA: National Council of Teachers of Mathematics, 2000.

Schoenfeld, A. H. "On Mathematics as Sense-Making: An Informal Attack on the Unfortunate Divorce of Formal and Informal Mathematics." In *Informal Reasoning and Education,* J. Voss, D. Perkins, and J. Segal (Editors), pp. 311–343. Hillsdale, NJ: Erlbaum, 1990.

The Emperor's Vanishing Clothes

Dan Kennedy
Baylor School, Chattanooga

One of the occupational hazards of teaching at an independent secondary school for more than a few years is that you are eventually asked to serve on a curriculum committee. I have actually been around long enough to have served on three. These committees, consisting of about a dozen teachers from all academic departments, are usually charged with the daunting task of reviewing graduation requirements in light of the changing needs of society. Predictable turf battles ensue, wherein each department attempts to convince the others that the citizens of the future need increased exposure to that department's courses, a premise that is rarely challenged until somebody suggests that this might occur at the expense of some other department's slice of the academic pie. Many heated meetings later, the frazzled committee members finally reach a compromise between change and tradition by resolving to tinker with the school's class schedule until everything can be added without anything appearing to have been subtracted. This necessitates several more months of meetings.

The amazing thing about these turf wars is that, while the sciences, arts, and social studies slug it out, English and mathematics remain seemingly above the fray—despite the fact that these are the only subjects that most students take every year. Nobody, apparently, dares to suggest that the need for knowledge in either of these classic disciplines will decrease in the future. In fact, our own department has often had to apologize for not making four years of mathematics a graduation requirement, a position we can afford to take simply because ninety-nine percent of our students are already electing a fourth year of mathematics at the urging of their parents and college counselors.

Contrast this unquestioning enthusiasm for mathematics in the curriculum with the widespread belief of so many people that they cannot "do" mathematics, or, indeed, with the bad memories that so many people seem to harbor about their own mathematics courses, and you are faced with a bit of a conundrum. Why are so many educated people so eager to visit upon their children what might reasonably be considered to be the mistakes of their past?

I submit that this irrational behavior derives from the fact that parents have not understood what we mathematics teachers have been teaching their children. The mere fact that the subject is called "mathematics" has enshrouded it in an intimidating cloak of mystery beneath which few people have cared to peer. They have seen that the world is increasingly reliant on technology and increasingly data driven, and they correctly perceive that some people understand these changes although they themselves do not. Blaming their own lack of understanding on their inability to do mathematics, and wishing their children to be among the future understanders, they conclude that their children had better learn as much mathematics as they possibly can, even if they themselves do not understand what it consists of. Thus it is that mathematics for so long has been given a free pass in curriculum committees.

People reading this response will probably recognize that what parents really want for their children is for them to become quantitatively literate. A parent reading "The Case for Quantitative Literacy" would, no doubt, nod enthusiastically at the lists of expressions and skills and say, "Yes! This is just what I want for my child!" That parent would have every right to assume that the mathematicians who created the need for this kind of literacy by now would have designed a curriculum enabling a child to learn it, especially since the expressions and skills involved in quantitative literacy can be so readily itemized.

Hiding in Plain View

Alas, this is not the case. Even in the most "reformed" of U.S. classrooms, students are being prepared for a capstone experience of college calculus and for embarrassingly little else. This obvious disconnect between supply and demand has endured only because we have been able to hide it behind the cloak of mystery that has enshrouded mathematics—the same cloak that has protected us in curriculum committees all these years. It is, for all its perversity, a cozy situation. But before we allow ourselves to become too comfortable, let me sound the warning to my colleagues in mathematics departments across the country that our cloak of mystery has started to unravel. If present trends continue, it is only a matter of time before our patrons see us standing at the chalkboard arrayed ignominiously in the emperor's new clothes.

As evidence, I offer several observations. First is the case statement, one of many recent documents calling attention to the contrast between the mathematics that ought to be learned and the mathematics that is actually being taught. These documents are not without precedent; indeed, the Mathematical Association of America's Committee on the Undergraduate Program in Mathematics (CUPM) has been urging curriculum reform for three decades, citing essentially the same computer-driven societal needs.

What makes the current reform agenda so powerful is that it has managed for the first time to reach *teachers* at every level, prompting a groundswell for change that has never before been present. Some mathematicians, hoping that the groundswell will pass, are trying to foment a "math war" in the hope of preserving the traditional curriculum. Unfortunately for their cause, there can be no war—because there is no army. Thanks to the changes already wrought by the standards for school mathematics, the calculus reform movement, and numerous exhortative documents such as the case statement, secondary school teachers who historically have been obedient foot soldiers on the precalculus drill team have now developed an appetite for relevant mathematics. So have many college teachers, and so have the university departments of mathematics education. The only math war that will attract their support now is that of addressing the challenges that the case statement presents. Most of my colleagues are not only willing, but also eager, to join that fight.

Second is the phenomenon of Advanced Placement (AP) Statistics. When the College Board introduced this course in 1995, it debuted with the largest opening volume of any AP examination ever. In five years the volume grew from 5,000 to more than 34,000, exceeding even the wildest predictions of anyone associated with AP. Lost in the dazzle of this unexpected growth has been the curious fact that most AP Statistics students are being taught by mathematics teachers who have never even taken a college statistics course. This apparent drawback has not prevented either the students or their teachers from loving the subject.

Statistics sessions now draw huge crowds at professional meetings of mathematics teachers, the Internet crackles with lively exchanges among statistics teachers from coast to coast, college departments are teaching more and more statistics courses, and editors of mathematics journals are suddenly hungry for articles about statistics. Many mathematics teachers (who are, after all, not engineers) are feeling for the first time in their careers the exhilaration of knowing firsthand what their upper-level courses are good for in the modern world. Understandably, they are dying to carry that enthusiasm into their algebra and geometry courses, but they are frustrated by the amount of traditional material that they still feel compelled to cover. In the parlance of the case statement, they really want to be teaching quantitative literacy to all their students.

Third is the catalyst that has brought us this far and that shows no sign of abating: technology. For better or worse, computer technology (for secondary school teachers, that means graphing calculators) is now inextricably entwined with mathematics education. The dramatic effect of this technology on the teaching and learning of mathematics since 1990 probably needs no recapitulation here. Let me simply note that the amazing capabilities of these machines have forced many of us to confront directly the questions of what algebra and geometry we ought to be teaching with the aid of technology and what should be taught without it. These questions have exposed the traditional mathematics curriculum to unprecedented scrutiny, raising along the way some embarrassing questions about its relevance. The case statement provides powerful evidence of why we can no longer avoid these questions.

Finally, the disturbing trend toward outright distrust of mathematics is perhaps the most apparent indication that the cloak of mystery is unraveling. The University of Rochester almost lost its graduate program in mathematics a few years ago because the administration had lost faith in its relevance. Many states, looking to hold secondary schools accountable for student performance in mathematics (and not content to let the teachers measure it), now require students to pass high-stakes tests administered by external organizations. Parents send letters to the editors of local newspapers protesting the adoption of particular mathematics textbooks in their districts. Education ministers in some countries (notably France and, sur-prisingly, Japan) have recently presented proposals for *less* mathematics in the curriculum.

It is apparent that many people are beginning to look more closely at the mathematics being taught in our schools, and we mathematics teach-ers have only ourselves to blame if parents and taxpayers do not like what they are seeing. Moreover, the noisy disagreement among professional mathematicians about what mathematics we ought to be teaching, far from being helpful, has surely increased the public's suspicion that we have nothing on underneath our vanishing cloak of mystery. After all, the only reason they trusted us when they did not know what we were talking about is because they thought that we, at least, did know. We need to jus-tify our courses to ourselves before we can justify them to our constituents.

One of the most common questions posed by students to their high school mathematics teachers is "What is this mathematics good for?" It is a question that we have been nervously answering for our 14-year-old algebra students throughout our careers. Fortunately for us, most of those youngsters can be taken in by a teacher's authoritative reply. The stakes are a lot higher now that we have to answer that question for impatient adults with clipboards.

We need better answers—and soon. We need to be teaching quantita-tive literacy.

Numerical Common Sense for All

Wade Ellis, Jr.
West Valley College

"We hold these truths to be self-evident, . . . "Our nation's call to revolution begins with a statement with a definite mathematical flavor. The remainder of the Declaration of Independence draws certain conclusions from these self-evident truths. Yet in illuminating what might flow from this foundation, the Constitution of the United States and its implementation severely limited opportunities for access to literacy and participation in the economy and politics of the country. For example:

- All men (but not women) were created equal.
- Women were not allowed to vote.
- Slaves were worth three-fifths of a free person in apportioning taxes.
- Slaves were not to be taught to read.

The documents on which the nation was founded assumed that logic could justify the revolution that they felt forced to undertake, and that basic literacy, if not universal, was sufficiently widespread to sustain the nation. In an agricultural society (at the time of the revolution nearly ninety percent of the U.S. population was involved in farming), this low level of literacy was sufficient. As the economy of the United States has moved from farm, to commercial, to industrial, to knowledge-based, the level of literacy required of the population has increased, as has the level of education provided by the state. By 1920, nearly universal junior high school education made the presumption of universal literacy common, although many still could not read. Only about four percent—the rich, well born, and lucky—pursued a college education, yet literacy in the

United States was considerably more widespread than in other countries. At the end of World War I, for example, the British Army had a literacy rate of less than fifty percent while the literacy rate in the United States was approximately seventy percent.

The increase in college and university enrollments that came with the end of World War II, aided by the GI Bill and driven by an expanding economy, resulted in a rapid increase in literacy in the United States. But what students were taught and what they could do with their knowledge began to diverge from the needs of society. School learning is just not the same as practical know-how. Historically, the United States has harbored a deep strain of anti-intellectualism: "If you're so smart, why ain't you rich?" Fortunately, this attitude has been somewhat balanced by a special appreciation for common sense and practical know-how: "Them engineers say that bees can't fly."

Common Sense for Today

In our complex, knowledge-based society, common sense now requires much more than the ability to read carefully and to think logically. The nation now requires that its citizens and workers have the ability to reason in a commonsense way in situations involving numbers, graphs, and symbols. Yet such abilities are de-emphasized and even shunned in our schools, colleges, and universities. The most important task now facing the education community is to find an appropriate balance between teaching rules of thumb that get the job done quickly and intellectual abstractions that eventually will create more effective and efficient rules of thumb.

Unfortunately, many teachers of algebra (myself included) provide instruction that constricts rather than expands student thinking. Several years ago I discovered that students leaving my elementary algebra course could solve fewer real-world problems after the course than they could before the course: they thought, after completing the course, that they had to use symbols to solve problems that they could previously solve with reasoning and arithmetic. Similarly, students in my differential equations course, although quite capable of manipulating symbols, often could not interpret the results of their manipulations either in the real world or in the purely mathematical world. Such educational failure reminds me of the Vietnam War stories of villages that had to be destroyed in order to be saved. Common sense says this was wrong, yet we continued destroying villages.

Politicians recognize that our schools are not producing students who can solve simple quantitative problems or use their mother wit to understand and cope with the world. It seems that we are not alone in this paradox. In Japan, a nation known for the high quality of student performance on international mathematics assessments, the academic community has decided to reduce by 30% the amount of time spent on mathematics in schools. Proponents of this change, including some of Japan's leading research mathematicians, argue that this reduction is appropriate because Japanese students, although they know a great deal, despise mathematics and are loathe to use it in their daily lives. A similar reduction is under discussion in Singapore as it tries to find ways of making mathematics more appealing.

These changes in countries known for their excellence in mathematics education suggest many challenges for the United States, a country not known for similar excellence. What educational opportunities should we provide for our students to increase their quantitative common sense so they can be more efficient workers and more effective members of the community? Who should be in charge of this effort to increase quantitative common sense or quantitative literacy? What, indeed, is quantitative literacy?

For me, quantitative literacy is more like art than science. I know it when I see it, but I cannot easily define it. On any given day, for any one person, quantitative literacy may include reconciling a bank statement, analyzing data to support or oppose a local government proposal, estimating how to split a lunch bill, debugging a program by working from assumptions toward a logical conclusion, deciding which medical treatment to pursue based on statistical evidence, building a logical court case, or understanding the risks in investing for retirement.

Fortunately, opportunities to use quantitative literacy abound. Unfortunately, few in the mathematics community see quantitative literacy as important enough to emphasize in their teaching. This is partially because we mathematicians have a different agenda for our students but also because we really do not know how to teach quantitative literacy effectively. This may sound like an insult but it really is not. We know what we know—mathematics—very well, but the quantitative literacy that students really need to learn touches on every aspect of life. For example, what percentage of people in the United States die alone in institutions and what does that mean for you and yours? Quantitative literacy is important because no one knows what life will be like in the future.

Quantitative Literacy in the Curriculum

Mathematicians need help to develop curricula that provide students opportunities to be involved both in abstract thought and practical problem solving. These two goals are not mutually exclusive but mutually supportive. The problem of quantitative literacy is not a deficiency that someone has to be blamed for but a symptom of the monumental changes our nation has experienced during its two centuries of existence. All nations, as they move into a knowledge-based environment, face similar problems and all will have to develop their own solutions that match the special needs of their populations.

   Academic mathematicians point out that the quality and quantity of mathematics graduate students who received their secondary and college mathematics education in the United States have reached dangerously low levels. Some suggest, in response, that we should change the entire secondary school curriculum to more clearly emphasize abstract symbolic reasoning, even if it is at the expense of real-world, data-driven analysis and problem solving; however, because we are only talking about a few hundred mathematics Ph.D.'s per year, common (quantitative) sense suggests that the benefits may not outweigh the cost of such a drastic change in pedagogical practice.

   Others say that applied real-world problems are so important that they should be taught in every discipline, even if at the expense of abstract pure mathematics. Clearly neither of these extremes will serve the country well. A balance is required and will eventually be reached. I believe in the multiplication tables (yes, through the 12s) and the distributive law, not only because they are needed to understand algebraic problem solving well enough to get correct answers but also because they are (in part) the basis of quantitative literacy—which encompasses citizenship, personal finance, personal medical decisions, and work-related spreadsheet analyses. As do many mathematicians, I embrace real-world problems (if carefully chosen and not overemphasized) because they can engage students in the abstraction, generalization, and logical thought that are the lifeblood of academic mathematics.

But even if mathematicians wanted to, they cannot teach quantitative literacy alone. Mathematicians need the help of other disciplines in secondary school and university departments to support the cause of quantitative literacy. Interdisciplinary activity is becoming more and more common and is beginning to be rewarded in academic life. Indeed, interdisciplinary work may now be possible without damaging a faculty member's chances for tenure or level of prestige. We need more courageous mathematicians (and other faculty) who will risk working outside the usual boundaries of the academic reward system. We need more courageous mathematicians who are willing to do both applied and pure mathematics even at the risk of criticism from their colleagues. Examples of such courage have begun to appear at the highest level in other countries. Many in this country have shown this courage as well.

The world of mathematics is changing, not only in practice but also in its very nature. It will continue to change, as everything does. Certainly it will always rely on the logical foundations of Euclid, the symbolic contributions of Vieta, the analytic contributions of Gauss, Cauchy, and Weierstrass, and on digital computing pioneered by von Neumann. Mathematics has a wonderful and exciting past. It is having a wonderful and exciting present. The future, however, is at risk because we in the mathematics community have not focused sufficiently on the needs of society. We can satisfy these needs in many ways—by preparing teachers, by offering ideas for developing quantitative literacy, and by using that quantitative literacy to encourage more students in the United States to become users of mathematics as well as mathematicians.

In the continuing quest for life, liberty, and the pursuit of happiness in our knowledge-based nation, we need to provide a new kind of common sense—a quantitative common sense—based on basic mathematical concepts, skills, and know how. The mathematics community will participate in providing our nation, both young and old, male and female, native born and immigrant, rich and poor, of every race and faith, with this new common sense. We will do it gladly, secure in the knowledge that we are building the intellectual foundation for our cherished discipline while supporting the well being and continued success of our nation.

Mathematics and Numeracy: Mutual Reinforcement

Alfred B. Manaster

University of California, San Diego

"The Case for Quantitative Literacy" offers a rich and diverse view of the numeracy needs of society, a view that encompasses much of mathematics and statistics. But the case statement also acknowledges that quantitative literacy has had and continues to have many meanings. In this response, I will focus on one particular aspect of quantitative literacy—the analysis of data—to draw some distinctions between quantitative literacy and mathematics. I believe these distinctions may facilitate a clearer understanding of what each area has to offer students and thus help educators give adequate emphasis to some of the special perspectives of each.

Working Definitions

The case statement makes a persuasive call for increased attention to numeracy in schools and colleges. In contrast, this response presents recommendations for mathematics education, some explicit, others implicit. Because quantitative literacy is so closely linked to school mathematics, those of us who teach mathematics need to consider very carefully the relationship between mathematics and quantitative literacy.

Mathematics is a well-known subject with a venerable history. A common dictionary defines "mathematics" as the science of (a) numbers and their operations, interrelations, combinations, generalizations, and abstractions and (b) space configurations and their structure, measurement, transformations, and generalizations. Numbers and shapes, arithmetic and geometry: these everyone recognizes as the essential foundation of mathematics.

In contrast, it appears from the case statement that the meaning of quantitative literacy is not well established. Indeed, the case statement devotes many pages to descriptions of elements, expressions, and skills that collectively describe quantitative literacy. Taking the words at their face value, we might summarize by saying that quantitative literacy is the ability to understand and reason with numerical information. That ability enables people to be comfortable with numerical data and to use them in meaningful ways, in particular to make well-reasoned decisions.

Using data to make decisions is rather different from the science of numbers and shapes. Yet even the latter well-accepted formulation has generated a wide variety of working definitions of mathematics among mathematicians, mathematics education researchers, and mathematics teachers, to say nothing of philosophers, engineers, scientists, and sociologists. So we should not be surprised when an emerging notion such as quantitative literacy produces the extraordinary variety of illustrations found in the case statement.

To compare mathematics with quantitative literacy, we need to focus not on the variety of definitions but on a few elements that are characteristic of each. For example, in mathematics, reasoning and proof play a central role. A critical feature of mathematics is understanding why assertions based on assumptions must be true. In contrast, numeracy frequently requires inferences based on estimates and approximations, on incomplete or sometimes inaccurate data.

Another distinctive characteristic of mathematics is that its assertions are about relationships among abstractions, whereas the inferences of quantitative literacy are almost always about something real. The first of these mathematical abstractions, both historically and pedagogically, is whole numbers. Other kinds of numbers are also studied, and these several number systems form the basis of many further abstractions, especially in algebra and analysis.

As the dictionary definition suggests, space configurations provide yet another source of abstractions that become objects of mathematical study. Other patterns can also serve as the basis for mathematical abstractions and assertions about their relationships. Quantitative literacy applies results from these mathematical abstractions to gain insight into concrete situations.

The Role of Numbers

Numbers, which are themselves abstractions, play important but different roles in mathematics and quantitative literacy. In quantitative literacy, numbers describe features of concrete situations that enhance our understanding. In mathematics, numbers are themselves the objects of study and lead to the discovery and exploration of even more abstract objects.

As the case statement makes clear, quantitative literacy offers people a quantifiable perspective for understanding the world. It does this primarily by reasoning with information obtained through measurements of quantifiable phenomena. The range of situations that can be quantified is very broad and constantly expanding. Familiar examples going back hundreds of years include the sizes of physical objects and the duration of events. More recent common examples include income distribution in populations, the differential effects of various medical treatments, and the relative popularity of different programs, policies, or politicians. In all these settings the role of numbers is descriptive: numbers help us understand and compare features of real-world situations and enable us to make decisions based partly on those features.

In mathematics, numbers play a very different role. The goal of a mathematical study of numbers is to obtain a better understanding of how numbers work within their own structures, of the properties they exhibit or lack. In contrast to quantitative literacy, in which numbers are descriptors of characteristics of the objects being studied, in mathematics numbers themselves are the objects of study.

When properties of specific numbers are studied, mathematicians quickly discover that it makes sense to look not at individual numbers but at systems of numbers and then to form even more abstract structures based on these systems. Reasoning about those general structures can provide insights that eventually may be applied to real-world problems, but mathematicians often are more interested in the abstract structures than in the concrete situations. Whereas the power of quantitative literacy derives from its faithfulness to real problems, the power of mathematics comes from its abstractness and consequent generality.

Historically, the need to measure objects and other phenomena provided a natural motivation for the introduction and development of numbers. The ways in which numbers are used to measure and compare objects and to analyze their properties are the first steps toward quantitative literacy. At the same time, learning how to compare numbers and combine them to form new numbers in meaningful ways are among the first steps in acquiring mathematical knowledge and ways of thinking.

Contrasts and Comparisons

From their common beginning in numbers, quantitative literacy and mathematics necessarily diverge, especially in school education. The elements and expressions enumerated in the case statement provide examples of these differences. This divergence of emphasis does not require completely separate instructional paths, but it does require that teachers give adequate and separate attention to the goals of each. At times, numbers and abstractions of number systems should be studied to understand and master efficient computational algorithms and to see what properties they satisfy. At other times, elements of this abstract knowledge need to be applied to understand models of complex concrete situations.

These different perspectives lead to different roles for the various types of reasoning employed in quantitative literacy and in mathematics. Although deductive, inductive, and analogic reasoning are important in both, the first seems more central to mathematics while the second and third play more prominent roles in mathematical modeling and quantitative literacy.

Mathematics provides many examples of knowledge that is important for understanding real-world phenomena, such as the order relation on real numbers, efficient computational algorithms, derivation of indirect dimensions from direct measurements, and use of scientific notation to express very large and very small numbers.

On the other hand, not all mathematical results, even about numbers, are directly applicable to or motivated by concrete situations. The important Fundamental Theorem of Arithmetic is a good example. It asserts that every integer greater than 1 is a unique product of prime numbers (except for reordering). Although this theorem turns out to have unexpected applications in logic and cryptography, these are neither the reasons it was discovered nor the primary reasons for learning why it is true.

Increased Demand

It is easy to understand the case statement's call for more emphasis on quantitative literacy. Technology has moved us into the information age. Technological improvements have broadly expanded both what can be measured and the accuracy of measurements. Increasingly sophisticated mathematical techniques convert complex measurements into tables containing massive amounts of data. These changes in turn confront decision makers with overwhelming amounts of data. The Internet, itself a product of technology, has made previously inconceivable quantities of data available to almost anyone who wants them. Understanding how these data are created—both technologically and mathematically—is an important prerequisite for being able to use data meaningfully.

This expanded availability of data has affected many areas of society. Fifty years ago most people who had knowledge of the natural sciences and engineering understood the fundamental importance of measurement and data. But there were very few such people. At that time access to huge quantities of statistical data was quite limited because of the unacceptably high costs of acquiring raw data and analyzing them to produce meaningful summaries.

Relatively inexpensive and widely distributed technology now enables the collection and analysis of massive amounts of data. One consequence is the dramatically increased role of quantitative data analysis in the social sciences. Beyond the social sciences, data are also used, for example, in political campaigns, to inform government decisions, and by businesses to develop product designs and marketing strategies. These and similar uses profoundly alter our cultural climate. Some understanding of the meaning and sense of data, including their acquisition and manipulation, is required for intelligent participation in a society in which decisions increasingly rely on interpretations of data.

In this sense, as the case statement argues, quantitative literacy provides an important complement to mathematics—the one emphasizing reasoning with data, the other reasoning with numbers and shapes. Although quantitative literacy and mathematics share a common root in numbers, they diverge in emphasis and subject matter. Yet they remain inextricably interconnected, especially in education. As students progress through their education, they should be given many opportunities to see how each supports the other.

Specifically, as students acquire mathematical sophistication they need opportunities to apply what they learn to the analysis of quantitative data and the understanding of real-world phenomena. Reciprocally, their study of increasingly complex natural and cultural phenomena should provide contexts for developing deeper mathematical understanding. The classic example of this reciprocity is velocity and acceleration, but analyses of opinions and behavior now are quite common as well. In this way a balanced educational program can enable numeracy and mathematics to reinforce each other.

Connecting Theory and Practice

An Interview with James H. Stith

American Institute of Physics

The case statement argues that quantitative literacy is not merely a euphemism for mathematics but is something significantly different—less formal and more intuitive, less abstract and more contextual, less symbolic and more concrete. Would quantitative literacy as thus described be particularly helpful to physics students, or are they best served by mastery of traditional mathematics?

The quantitative literacy described in the case statement would serve as an outstanding foundation for all students, regardless of their potential major in college or their avocation should they decide to directly enter the workforce. Although the traditional mathematics curriculum has served the physics community well, I don't believe that quantitative literacy as defined in the case statement precludes a student's mastery of traditional mathematics. Furthermore, there is a real possibility that quantitative literacy as here defined might increase the yield of physics students from the pool of those who currently take the introductory course (about 1 in 32).

For example, quantitative literacy might give students a better understanding of the difference between change and rates of change, leading to a better conceptual grasp of the difference between velocity and acceleration. In addition, QL might sharpen the proportional reasoning skills of students, leading to a better grasp of density. Many students struggle to understand the differences between linear density, surface density, and volume density, often using the three quantities interchangeably. Students' inability to build a conceptual model of these ideas is reflected in their difficulty understanding and applying follow-on principles. Far too often, many of these students drop out of the program, never taking another course in physics.

Too many students (and instructors) link students' difficulty in the physical sciences to their lack of understanding of mathematical concepts. Far too frequently, faculties tie an understanding of physical science concepts to the ability to manipulate mathematical equations. Quantitative literacy as defined in the case statement could assist in changing this status quo.

> *Mathematics is indeed the culprit in many of the problems students have in science. It is a subject students love to hate. Can an emphasis on quantitative literacy help resolve this age-old dilemma? Should schools give greater emphasis to quantitative literacy than to the more formal aspects of mathematics?*

The question is, Why do students love to hate mathematics? If the answer is that too many mathematics teachers hide behind the formalism of the subject, then emphasis on quantitative literacy may well help resolve the dilemma. Given the mounting evidence that students learn better when they can relate to a subject, when they can see the "big picture," it makes sense that a greater emphasis on quantitative literacy in the early years will help students understand and relate to the more formal aspects of mathematics when they come along later.

Furthermore, I believe that an emphasis on quantitative literacy will appeal to a broader cross section of the student population, thus giving them a better understanding of mathematical concepts while not eroding the level of understanding of those who thrive under the current system. I see QL not as watering down or otherwise changing the basic content of the subject but rather as raising the expectations we have for all students. Ideally, both faculty and students would believe that students can learn, understand, and use mathematics. An emphasis on quantitative literacy may help reduce the widespread perception that some people are born to do mathematics while others are not.

> *That's a very important point. Too many people see mathematical ability as an inborn characteristic that determines students' future*

options. Indeed, mathematics has been called a "critical filter" that blocks students with weak backgrounds from rewarding careers. Just how important is it that all students master formal mathematics? Might context-rich quantitative literacy be a more reasonable expectation? Does it make sense for students with different interests, say in grades 10–14, to study significantly different kinds of mathematics?

I have difficulty drawing the distinction between formal mathematics and context-rich quantitative literacy as an either-or scenario. I think we would all agree there exists a subset of formal mathematics that all students, regardless of what they choose to do in life, should master. I believe we would also agree that there are certain habits of mind that we want all students to exhibit.

It does not make sense to me that students in grades 10–12 should study significantly different kinds of mathematics, but it does make sense that those in grades 13–14 perhaps should do so. My feeling is that fundamental algebraic, geometric, trigonometric, and probabilistic concepts coupled with a strong sense of logic will stand all students in good stead and provide the support that all subjects requiring a mathematics foundation can and should build on.

I would reiterate, though, that the "what" that is taught is not nearly as important as how that material is taught. I don't underestimate how difficult it will be to change the habits, beliefs, and pedagogical practices of a significant fraction of the teaching workforce. But a fundamental change must occur to reach the point where everyone truly believes that all students can learn mathematics and science. The question is, How do we move beyond the rhetoric?

Suppose for the moment that we have moved beyond rhetoric and that all students study the same core in high school. What level of quantitative literacy should be required to enroll in a four-year degree program? How much more, if any, should be required for all bachelor's degree recipients?

Every student enrolling in a four-year degree program should have a basic understanding of the arithmetic, geometric, algebraic, and trigonometric concepts taught in most, if not all, high school mathematics programs. Additionally, every college graduate should have a fundamental understanding of statistical and probabilistic concepts. Students should be required to demonstrate their understanding of those concepts by using them in a variety of real-life situations.

This sounds pretty much like requiring the traditional precalcu-lus mathematics sequence for everyone who plans to go to college. Yet the case statement suggests a rather different vision of high school mathematics. Help me understand how you see the balance between traditional mathematics and quantitative literacy playing out in the high school curriculum. How can both aims be achieved at the same time?

Maybe I read the case statement incorrectly, but I did not see it call-ing for fundamental changes in content itself. Rather, I saw it calling for moving beyond content. My belief is that the traditional precalculus math-ematics sequence suggested for those who plan to go to college will serve all students well and hence should be required of all students. I agree that the high school curriculum is short on statistical concepts. Although it would be desirable to introduce those concepts in high school, it is essen-tial that they be part of every college graduate's portfolio.

We need, however, to move beyond the goal of teaching students a set of skills that is nice to have to teaching a set of competencies that students feel comfortable using on a daily basis. Students should see their mathe-matical skills as tools that are important to use on a daily basis. They should be as comfortable with the language of mathematics as with the language of English. The threads of mathematics should be present and exploited in every course in the high school curriculum. Logical thinking is as important in history and social science as it is in mathematics and science. All mathematics taught in high school should be seen as "mathe-matics for life" for all students. College preparatory mathematics has, I believe, become a misnomer. Hence, I really resonated with the "Ele-ments of Quantitative Literacy" section of the case statement.

Some critics worry that a focus on quantitative or scientific liter-acy weakens students' backgrounds by providing only vicarious experiences—a distant observer's knowledge about science or math-ematics. Do you see a distinction between traditional lab-based science and science literacy, or between traditional skills-based

mathematics and quantitative literacy? Which is more important for
students in grades 10–14?

If done well, I don't believe that a focus on quantitative or scientific literacy weakens students' backgrounds. On the contrary, students' scientific and mathematical understanding and ability to extend what they have learned to the next level is only enhanced.

I have always felt that teaching a conceptual physics course is far more difficult than teaching the so-called traditional physics course that is steeped in mathematics. Far too often, the instructor writes an equation on the blackboard that embodies the essence of a physical law without ever getting the student to see the subtleties that are clear to the instructor. What makes matters worse is that often neither the student nor the instructor recognizes that there has been a communication gap. This same gap appears between traditional skills-based mathematics and quantitative literacy. Hence our most important goal is quantitative literacy for all students.

Is it reasonable to teach quantitative literacy "across the curricu-
lum" as writing often is taught, or does it require special expertise?
Or might it be the case that science teachers in fact are better able to
teach quantitative literacy because they, in contrast to mathematics
teachers, also deal with real contexts in which quantitative issues
arise?

Ideally, it would be great to teach quantitative literacy across the curriculum, for in my view this is the only way that students can see the connection between school mathematics and the mathematics of real life. It is not clear, however, that science teachers hold any special edge when it comes to the ability to provide the context for the use of mathematics within a particular subject.

I have some real concerns about the communication gap between science and mathematics teachers that often leads to student confusion, such as when similar quantities are called by different names in different subjects. The silo structure that separates disciplines at the college and graduate school levels unfortunately perpetuates itself at the school level.

I would claim that mathematics teachers, like all teachers, deal on a daily basis with the real context in which quantitative issues arise. Yet for some reason, the paradigm is that in school mathematics this context is often removed. The result is that students often walk out of the classroom making a distinction between theory and practice when they should really be seeing the connections between theory and practice.

Why is context so often removed from school mathematics? We constantly hear stories of students who learn mathematical formalities without any accompanying sense of meaning. Is this a problem of pedagogy, or is it perhaps inherent in the nature of mathematics? Does it also happen in physics?

To answer the last part of the question first, yes, this also happens in physics. My belief is that in both subjects, this disjunction is not inherent in the nature of the subject but is a problem of pedagogy. Although there is a growing body of knowledge on how students learn, it is my sense that most faculty are unaware of this research and hence do not incorporate it into their daily teaching practice.

My sense is that most instructors teach their subject using techniques they believe make the material most understandable. Unfortunately, they frequently use younger versions of themselves as the model for their students and expect their students to supply many of the missing steps, just as they did when they were students. Moreover, many instructors do not remember the difficulty they had mastering these same concepts when they were first exposed to them. The end result is that much of the context is eliminated. Finally, putting in the context is often perceived as taking time and is thus ignored so that more "content" may be covered. We must somehow overcome the deep-seated conviction that content is king, that if we can get the content correct, the rest will take care of itself.

Quantitative Literacy
for the Next Generation

Zalman Usiskin

University of Chicago

"The Case for Quantitative Literacy" is a well-informed, well-documented, and persuasive statement on the importance of quantitative literacy. When we see an argument made that adults need to know something that evidence shows they do not know, we naturally turn to the schools for the answer. Accordingly, the case statement concludes that educational policy should promote a strong emphasis on quantitative literacy. But before an already full school curriculum takes on a new concern, we must ask if this concern has been addressed previously. If quantitative literacy already has been addressed, perhaps it is more difficult to learn than we thought. Or we may think we are addressing quantitative literacy when we are not, or not doing it well or well enough. If quantitative literacy has not been addressed to the extent we would like, we need to suggest specific steps that might be taken by teachers and curriculum developers to remedy the situation.

Few of the contexts described in the first paragraphs of the case statement—increases in gasoline prices, changes in SAT scores, risks of dying from colon cancer, numbers of refugees, costs of cell phone contracts, interest on car loans, team statistics in sports, odds in competitions, locating markets in business, analyzing soil, measuring drug dosages, etc.—are ever mentioned, let alone studied, in the mainstream school mathematics curriculum. Most of these contexts, however, did not even exist a hundred years ago when the school curriculum was, for all practical purposes, codified. Quantitative literacy today thus involves many applications that are so new (relative to the pace of educational change) that the school mathematics curriculum has not caught up with them.

Schools should not engage in self-flagellation for being behind current uses of mathematics. Continual updating is a necessary circumstance in such a vibrant subject. School mathematics cannot be expected to foresee new applications of mathematics; it must lag behind them. But we should be concerned if the adults of tomorrow do not possess the quantitative literacy that we know is needed today.

Most of the adults who do possess quantitative literacy learned it outside the mathematics classroom in much the same way they learned about computers. They developed some number sense, arithmetical confidence, and experience with data by being interested in sports and calculating sports statistics, by working in a store and dealing with money in and money out on a regular basis, by handling their own taxes and financial affairs, by building models, or by engaging in any of a large number of other activities. They could not have learned quantitative literacy in the mathematics classroom because historically the mainstream mathematics curriculum has devoted extraordinarily little attention to the connections of mathematics with the real world.

Yet today a large number of adults do not possess quantitative literacy. Whatever these adults may have learned in school, it did not include the prerequisites for quantitative literacy. Although much of the standard curriculum certainly has the potential to enable students to gain number sense, practical skills, and symbol sense (three of the aspects of quantitative literacy discussed in the case statement), mathematics as it has been taught in most schools does not help individuals gain quantitative literacy. In fact, some aspects of the school mathematics curriculum have worked *against* the development of quantitative literacy. If we want the next generation to be quantitatively literate, we need to avoid these aspects by adopting at least the following five practices in school mathematics classrooms:

- *Spend less time on small whole numbers.*

Despite the perceptions of some education critics, the understanding of small whole numbers in the U.S. population is very high. National Assessment of Educational Progress (NAEP) studies show that virtually all 17-year-olds in school compute accurately with small whole numbers and can apply operations on them appropriately in simple situations (Campbell, Voelkl, and Donahue, 1997, 57).

The reason for such stellar performance is the intense amount of time spent on these numbers and very little else. Although research indicates that most children enter kindergarten knowing how to count well past 20, traditional kindergarten mathematics programs spend most of the time counting to 10. By first grade most children can count past 100 (clearly learned outside of school), but first-grade texts spend a majority of the time on numbers no larger than 20. This slow start cannot help children develop the ability to deal with larger numbers.

Furthermore, the lower numbers are dealt with concretely, with models for counting, but when the larger numbers appear, they are almost always presented symbolically in the context of computation. In landmark studies in the second quarter of the twentieth century, William Brownell showed that teaching with attention to properties that relate numbers to each other (what he called *meaningful learning*) leads to higher performance than treating number facts as isolated bits of knowledge (*rote learning*) (Brownell, 1935). Adults' lack of ability to deal with arithmetic, even when computation was virtually the entire arithmetic curriculum, indicates that rote paper-and-pencil computational practice and facility does not enhance understanding of "number."

Those of us who have spent time teaching the real-world uses of mathematics believe that if we want people to understand these real contexts, we must teach them specifically and not expect automatic transfer from theoretical properties and relationships to practical situations. Contexts that apply fractions, decimals, percentages, and large numbers can fulfill the role of concrete experience for many students and lead to greater understanding of both theory and practice.

- *Give meanings of arithmetic operations that apply to numbers other than whole numbers.*

What does subtraction *mean*? An adult is likely to say "take away." Ask an elementary school teacher what multiplication *is*. If any answer at all is given, it is likely to be "repeated addition," for that is the meaning given in most elementary school textbooks. In books, division means splitting up into equal portions or repeated subtraction. Taking powers means repeated multiplication.

Each of these meanings applies well to whole numbers but each fails for fractions, decimals, percentages, and negative numbers. "Take away" does not explain why a temperature of 6° is 10° higher than one of -4°. Repeated addition does not readily explain why we multiply 9.25 by 12.417 (or 9¼ by 12⁵⁄₁₂) to find the area of a rectangular room 9' 3" by 12' 5". Nor can splitting up or repeated subtraction easily explain why we divide to determine the speed of a runner who has run 100 meters in 11.35 seconds. Similarly, repeated multiplication does not explain the calculation $500(1.06)^{2.5}$ that gives the amount to which $500 will grow in two-and-a-half years at an annual percentage rate of 6%.

Because of the ubiquity of calculators, applications of arithmetic no longer require that students be able to obtain the answers to the above questions by paper and pencil. These and many other applications do require, however, that the user understand the fundamental connections of subtraction with comparison, multiplication with area (and volume), division with rate, and powers with growth.

- *Employ calculators and computers as a natural part of the curriculum.*

I am old enough to remember when the first hand-held calculators appeared. I recall my immediate reaction: a godsend that would finally free us from the shackles of paper-and-pencil arithmetic. Thirty years later my belief is even stronger because calculators do far more than replace tedious arithmetic. Current technology enables ordinary people to work with mathematics that hitherto was inaccessible because of the computational limitations of paper and pencil.

Furthermore, current technology has caused much of the increase in the need for quantitative literacy. Without this technology, newspapers, financial institutions, scientific endeavors, and everything else that uses mathematics would not be the same. Why can a sports section report changes in baseball batting averages along with the box score from the game? Why can the price/earnings ratios of thousands of stocks be reported daily? What gives online financial planning sites the ability to calculate expected returns from any number of retirement plans in just seconds? The answer is the ability to do *automatic calculation*, calculation programmed into a spreadsheet or computer.

Applying these and a myriad of other uses of number in quantitative situations does not require that the reader or recipient be able to duplicate the calculator or computer. But it does require the ability to understand what the results of the calculations mean and to verify whether the calculations are correct to within some bounds of correctness.

• *Combine measurement, probability, and statistics with arithmetic.*

The content of the current school mathematics curriculum is often split into five strands: number (including computation), measurement, geometry, algebra (including functions), and probability and statistics. These are the strands by which items on the mathematics portion of the National Assessment of Educational Progress are classified (National Mathematics Consensus Project, undated); they are also the five content strands of the *Principles and Standards for School Mathematics* developed by the National Council of Teachers of Mathematics (NCTM, 2000).

It could be argued that any split of this type is artificial because all these strands are related; however, the separation of number from measurement has particularly negative implications for the teaching of quantitative literacy. For example, perimeter and addition, area and multiplication, similarity and division, and measurement conversion, multiplication, and division are inextricably related in reality, but rarely in school.

The study of measurement in elementary mathematics textbooks is almost exclusively devoted to geometric measures. Money, the most familiar form of measure to students, which could be used as an example of unit conversion both within and between systems, is not treated as measurement. Units of time are not studied along with other units. Units other than of length or weight (or mass) are rarely seen. Even counting units are not treated as units. Students most often encounter numbers as decontextualized.

Likewise, textbooks treat probability as a subject separate from arithmetic, unrelated to division even though probability is typically defined as a ratio. In the sections of the textbooks devoted to probability, there are discussions of a 25% chance of an event occurring, or equivalently that there is a 1 in 4 chance of occurrence, but these sections are often skipped by teachers. Probability is seldom mentioned in the sections on fractions and percentages that are always covered by teachers.

These artificial separations also apply to statistics. Students are not introduced to statistics as measures of a sample or a population, but as numbers resulting from formulas that typically rise from out of the blue. For instance, students do not learn to relate averages to division. They do not study percentiles as relating to percentages. Also, little attention is given to the origins of the numbers used in the calculations. A sample value is viewed as the value of a population without thought given either to the nature of the sample or the population it does or does not represent.

Students seldom study scales or the normal distribution, so they have little idea of the distributions of numbers that can be pivotal in their school lives: grade-level scores on standardized tests or scores on the SAT and ACT college admission tests. The ignorance of previous generations in these areas leads to major errors in the interpretation of test outcomes, such as viewing test scores as fixed rather than as sample measures subject to variability. This results in students being included in or excluded from academic programs based on differences in test scores that are too small to be significant.

- *Do not treat word problems that are not applications as if they were applications.*

The many examples of the need for quantitative literacy offered in the case statement can easily lead us to wonder why so little has been accomplished. I believe the problem relates in part to a perception by the majority of mathematics teachers about the "word problems" or "story problems" they studied in high school (e.g., "Mary is half as old as her father was . . ." or "Two trains leave the station one hour apart . . ." or "I have 20 coins in my pocket, some dimes . . ."). These problems have little to do with real situations and they invoke fear and avoidance in many students. So it should come as no surprise that current teachers imagine that "applications" are as artificial as the word problems they encountered as students, and feel that mathematics beyond simple arithmetic has few real applications.

I do not mean that the standard word problems should not be taught, or that there is no room for fantasy in mathematics. Most of the typical word problems make for fine puzzles, and many students and teachers appreciate this kind of mathematical fantasy. The point is that these kinds of problems are not applications, nor should they substitute for them.

This historical argument suggests that quantitative literacy will not become mainstream in our schools until a generation of teachers has learned its mathematics with attention to quantitative literacy—a chicken-and-egg dilemma similar to that regarding the public apathy about quantitative literacy described in the case statement. Short of forcing teachers to teach in a particular way—a practice I oppose on democratic grounds—there are at least two ways this dilemma can be resolved. We can hope that tomorrow's teachers will have studied in or are teaching one of the recent curricula that pay attention to the various aspects of quantitative literacy. Or we can engage in massive teacher training in quantitative literacy.

We may be able to obtain public support for attention to quantitative literacy if we emphasize that quantitative literacy is an essential part of literacy itself. From the first page through the feature articles, advertisements, editorial pages, business, entertainment, and sports sections, newspapers are filled with numbers. The median number of numbers on a full-length newspaper page is almost always well over 100; the mean is over 500 (Usiskin 1994, 1996). Numbers permeate tabloids and magazines as well. The popular press is not known for its familiarity with mathematics, and its editors are unlikely to have majored in science or mathematics in college, but numbers cannot be avoided. Without quantitative literacy, people cannot fully understand what is in the everyday news, what is in everyday life.

REFERENCES

Brownell, William A. "Psychological Considerations in the Learning and the Teaching of Arithmetic." In *The Teaching of Arithmetic,* Tenth Yearbook of the National Council of Teachers of Mathematics, pp. 1-31. Washington, DC: National Council of Teachers of Mathematics, 1935; reprinted in *The Place of Meaning in Mathematics Instruction: Selected Theoretical Papers of William A. Brownell,* J. Fred Weaver and Jeremy Kilpatrick (Editors), Studies in Mathematics, Volume XXI, Stanford, CA: School Mathematics Study Group, 1972.

Campbell, Jay R., Voelkl, Kristin E., and Donahue, Patricia L. *NAEP 1996 Trends in Academic Progress.* Washington, DC: U.S. Department of Education, 1997.

National Council of Teachers of Mathematics (NCTM). *Principles and Standards for School Mathematics.* Reston, VA: National Council of Teachers of Mathematics, 2000.

National Mathematics Consensus Project. *Mathematics Framework for the 1996 National Assessment of Educational Progress.* Washington, DC: National Assessment Governing Board, U.S. Department of Education, undated.

Usiskin, Zalman. "From 'Mathematics for Some' to 'Mathematics for All.' " In *Selected Lectures from the 7th International Congress on Mathematical Education,* David F. Robitaille, David H. Wheeler, and Carolyn Kieran (Editors), Sainte Foy, Quebec: Les Presses de L'Université Laval, 1994.

Usiskin, Zalman. "Mathematics as a Language." In *Communication in Mathematics, K-12 and Beyond,* Portia C. Elliott and Margaret J. Kenney (Editors), Reston, VA: National Council of Teachers of Mathematics, 1996.

Encouraging Progressive Pedagogy

Larry Cuban

Stanford University

Should a high school graduate be able to reconcile a bank statement and locate where mistakes occurred? Understand that clusters of cancer occurrences in one city can occur by chance? Recognize that opinion poll results can be biased by poor wording of questions and the sample that pollsters used?

To reformers eager for high school and college graduates to be quantitatively literate (or functionally numerate), the answer is an emphatic yes. After all, although numbers are deeply embedded in our daily lives, the available evidence indicates low levels of numeracy among U.S. high school and even college graduates. From making sense of the school district's most recent test scores published in the daily paper, to parsing each presidential candidate's statistics on using the budget surplus to strengthen Social Security, to putting into plain words the doctor's estimate of your father's chances of surviving the spread of prostate cancer, numbers are everywhere.

Not mere background noise, numbers demand active sense-making. Just as verbal literacy gives students the tools to think for themselves, to question experts, and to make civic decisions, quantitative literacy does exactly the same in a world increasingly drenched in charts, graphs, and data.

During the past two decades of intense and sustained school reform led by an alliance of corporate executives, educators, and public officials from both political parties, two points have become increasingly clear. First, ever since 1983 when *A Nation at Risk* was published, the agenda of educational reform has concentrated on ensuring that public schools prepare workers for a highly competitive, information-based global economy. The prevailing strategy has been to raise graduation requirements by adding more mathematics and science, establish uniform curriculum standards for all students, create performance standards to measure subject-matter proficiency through standardized tests, and, most recently, to hold students, teachers, and principals personally responsible for achieving benchmarks on national tests.

As a result, high school graduates in 2000 took more mathematics and science courses, did more mathematics and science homework, and read from "better" mathematics and science textbooks than did their forebears. Today's teachers who are certified to teach mathematics and science are familiar with the new mathematics and science curriculum standards that began appearing in the late 1980s. And test scores have improved on national and international standardized tests in mathematics and science, although not to the degree desired.

Second, this reform agenda of binding public schools to the nation's economy has led inexorably to producing traditional schools and classrooms that in decorum, subject matter, and teaching style would make the grandparents of today's students feel at home. Within this overall climate of heightened concern for preparing students for college and information-based workplaces and increased emphasis on the newest technologies, mathematics and science teachers still lecture, require students to take notes, assign homework from texts, and give multiple-choice tests. If anything, in the past few years mathematics and science classrooms, while awash in graphing calculators and computers, have largely experienced a resurgence of traditional ways of teaching and learning. Those reformers who believed that students should take more mathematics and science and be held accountable for what they learned in these courses have certainly had their wishes fulfilled.

Two Challenges to the Case Statement

In making its compelling case for functional numeracy, however, "The Case for Quantitative Literacy" acknowledges that more mathematics and science will not automatically lead to numeracy. Quantitative illiteracy cannot be overcome by introducing more subject matter. As wise as their arguments are and as passionately as they believe in the importance of numeracy, the authors still overlook the basic historical lessons to be drawn from earlier reforms in curriculum and pedagogy, especially those of the past two decades. I offer just two lessons that the case statement neglects to consider.

Lesson 1: Curriculum and pedagogy are inseparable. If anything has been established in the history of teaching, it is the simple fact that a teacher's knowledge of content seldom guarantees that he or she can structure and communicate that knowledge in ways that enable a diversity of learners (particularly those who are compelled to attend classes) to understand and apply the knowledge that has been learned. How teachers teach matters. Policymakers who add subjects to the curriculum or extend the amount of time that students spend on subjects are, at best, establishing prior conditions for learning, not learning itself. Pedagogy, the art and science of teaching, is as essential to learning as fuel is to moving a car.

Lesson 2: The quest for numeracy is a plea for progressive pedagogy in schools. Historically, tensions between traditional and progressive views of teaching and learning have pervaded every subject in the curriculum. The history of mathematics, science, reading, writing, social studies, English, and foreign language as subjects taught in public schools has oscillated between pedagogical reforms under which teaching the subject depended heavily on traditional direct methods of instructions (e.g., lectures, question-and-answer recitations, careful reading of text, frequent tests on content) and progressive methods (e.g., connecting content to real-life situations, lighter coverage of topics, an emphasis on understanding concepts rather than facts, integrating content across disciplinary boundaries).

If the standards issued by the National Council of Teachers of Mathematics (NCTM) in the late 1980s (NCTM, 1989) urged on teachers a progressive pedagogy in mathematics teaching and assessment, both the recent revision of those standards (NCTM, 2000), which placed increased stress on basic arithmetic and accuracy, and the current climate signal that a shift in classrooms toward traditional methods has occurred and is more acceptable now. The case statement is an unabashed and persuasive lawyer's brief advocating more progressive approaches at a time when current reform agendas emphasize traditional pedagogy.

Implications

What are the implications of these two lessons for the case statement? In the past two decades the academic curriculum has been vocationalized; that is, more academics, more tests, more accountability have targeted only one goal of public schools equipping high school graduates for the workplace. Accompanying this emphasis has been a strengthening of traditional structures of schooling and classroom pedagogy. For me, the most obvious implication is that those who advocate quantitative literacy need to divorce themselves from the alliance that has vocationalized public schools and align themselves with other reformers who want more from their public schools than preparing workers and who understand that the prevailing structures of schooling influence how teachers teach.

Such school reformers, for example, see the primary goal of tax-supported public schools as nourishing civic virtue and participation in democratic institutions. They seek deep changes in how schools are organized and governed to make them consistent with the broader purposes of schools in a democracy. Proponents of numeracy need to join those civic-minded and pedagogical reformers who call for tighter connections between formal schooling and life experiences. They need to form alliances with teachers and administrators who want to restructure schools to make teaching consistent with quantitative literacy.

What the authors of the case statement offer is a sophisticated call for a more progressive pedagogy. All of the astute examples offered speak to the pervasiveness of numeracy in our daily lives rather than the compartmentalization of numbers into academic subjects. The statement makes a strong case for the relevance of teachers' expertise in choosing what content and which approaches will get students to learn. The statement also argues for broader and different reforms than those now popular. Without recognizing explicitly that this passionate call for quantitative literacy runs counter to the present direction in school reform, the authors risk having the statement become just another document that garners a few news stories when it is published and then disappears until a doctoral student, years later, footnotes their valiant position statement in a dissertation.

The rationale for quantitative literacy is compelling. I am convinced that it is crucial for U. S. schools and for a democratic citizenry fulfilling its civic duties. Those who drafted this statement need to be clear that their call for numeracy is a call for a different, more salient pedagogy than now exists. Once this is made explicit, the authors can then begin to mobilize educators and citizens who share their ideals.

REFERENCES

National Council of Teachers of Mathematics. *Curriculum and Evaluation Standards for School Mathematics*. Reston, VA: National Council of Teachers of Mathematics, 1989.

National Council of Teachers of Mathematics. *Principles and Standards for School Mathematics*. Reston, VA: National Council of Teachers of Mathematics, 2000.

Achieving Numeracy:
The Challenge of Implementation

Deborah Hughes-Hallett
University of Arizona

"The Case for Quantitative Literacy" makes a convincing argument that quantitative literacy is necessary in the contemporary world. It suggests the need for a campaign to increase quantitative literacy. It makes some interesting points about the differences between quantitative literacy and mathematics. But it says little about how quantitative literacy is to be developed. There are some hints—"throughout the curriculum"—but few specifics. There are some warnings—more years of high school mathematics or more rigorous graduation standards will not work—but no prescriptions or recommendations.

In this response, I want to suggest some strategies for developing quantitative literacy. But first, we need to explore the differences between teaching quantitative literacy and teaching mathematics to understand why traditional schooling has not led to quantitative literacy. This analysis suggests an approach to teaching quantitative literacy that can generate recommendations for the next steps in the campaign.

Quantitative Literacy, Mathematics, and Statistics

The case statement makes two points about the difference between quantitative literacy and mathematical knowledge. The first is essentially a statistical one, namely, that more mathematics course work (calculus, trigonometry, etc.) in school has not lead to an increase in quantitative literacy. A careful study likely would show some correlation between mathematical achievement and quantitative literacy because some mathematical skills are a necessary part of quantitative literacy. But there are many examples of students with sophisticated mathematics course work in their backgrounds who possess minimal quantitative literacy, as well as many examples of students with remarkable levels of quantitative literacy but little formal mathematics. This suggests that trying to improve quantitative literacy by requiring more mathematics courses is at best inefficient.

The second point made in the case statement is that mathematics focuses on climbing the ladder of abstraction, while quantitative literacy clings to context. Mathematics asks students to rise above context, while quantitative literacy asks students to stay in context. Mathematics is about general principles that can be applied in a range of contexts; quantitative literacy is about seeing every context through a quantitative lens.

Learning mathematics generally involves two steps: learning mathematical principles and identifying mathematics in a context. Although students find the first step hard, they find the second step harder still. (Hence "word problems" evoke such panic.) School mathematics focuses almost exclusively on the first step, however, while quantitative literacy hinges on the second. Thus, although it is possible to imagine an educational system in which more mathematics courses lead to an increase in quantitative literacy, we do not currently have such a system.

As the case statement points out, statistics is the quantitative tool most likely to be encountered by ordinary individuals, leading to the conclusion that statistics is closer to quantitative literacy than is traditional school mathematics. Both the American Statistical Association (ASA) and the National Council of Teachers of Mathematics (NCTM) have worked to include more exploratory data analysis in the school curriculum. The 1989 and 2000 NCTM standards envision the school curriculum as a bridge to statistics in the way that it traditionally has been a bridge to calculus. Thus it is reasonable to believe that curricula based on the NCTM standards are likely to be better promoters of quantitative literacy.

Teaching in Context

If quantitative literacy is the ability to identify quantitative relationships in a range of contexts, it must be taught in context. Thus, quantitative literacy is everyone's responsibility. There are opportunities to teach it throughout the curriculum; however, experience teaching applications of mathematics suggests that teaching in context may be difficult. Let us consider why this may be so.

Teachers of mathematics and science often complain that students have difficulty applying the mathematics they have learned in another context. Part of the reason lies in the way that subjects are taught—separate from one another. But part of the problem lies deeper. Recognizing mathematics in another field requires understanding the context. A student's ability to understand a context depends heavily on the relationship of that student to the field in question. For example, nonscience students in calculus often dislike applications from physics because they do not understand that field. On the other hand, the same students may easily grasp applications from the life sciences and economics. Experienced teachers are aware of this phenomenon and often tailor their examples to the class.

Some years ago I did an experiment to illustrate the effect of context on students at the algebra-trigonometry level. I created two sets of mathematically identical problems in two different contexts, one everyday (e.g., the distance to grandma's home) and one scientific (e.g., the distance between atoms). No scientific knowledge was necessary to do the problems and they were not complicated. For example, one problem involved using the Pythagorean Theorem to find the third side of a right triangle. Students first did the everyday problems and then the scientific set. As they worked, they recorded all their thoughts on audiotape.

The results were not surprising, although their vividness was stunning. As expected, students did not realize that the two sets of problems were fundamentally the same. In numerous instances, a student could do a problem on the first set but not the corresponding problem on the second set. What was unexpected was the vehemence with which the students reacted to the scientific context. The tapes provided an opportunity to observe the mechanism by which the context became a barrier. Students went off on extended tangents (about how much they hated chemistry, for example) that completely distracted them from focusing on the problem. They literally wore themselves out by unnecessary efforts—sometimes 20 minutes in length—that had nothing to do with the problem, and then gave up. And this was on problems that they had essentially already solved on the previous set.

On another occasion, I observed a group of students making heavy weather of their homework on the blackboard because it used the cumbersome and unfamiliar symbols "fish" and "fish-on-nose." * They too would have had a much easier time if the context—in this case the letters—had been familiar.

Students' mathematical common sense and ability to apply their knowledge are clearly fragile. They easily evaporate in an unfamiliar context. They appear to be more fragile than students' knowledge of mathematical algorithms. In my opinion, the observation in the case statement that "this inextricable link to reality makes quantitative reasoning every bit as challenging and rigorous as mathematical reasoning" significantly understates the difficulty many students have with quantitative reasoning. Students are not quantitatively literate both because quantitative literacy is not widely taught and because they find it hard. Thus, achieving quantitative literacy is an enormous challenge.

The Role of Insight

The reason that quantitative literacy is hard to learn and hard to teach is that it involves insight as well as algorithms. Some algorithms are of course necessary—it is difficult to do much analysis without knowing arithmetic, for example. But algorithms are not enough; insight is necessary as well.

Insight connotes an understanding of quantitative relationships and the ability to identify those relationships in an unfamiliar context. For example, a caller to *Talk of the Nation* (PBS, 2000) demonstrated insight when he pointed out that a tax of £8 out of £10 spent on gas is a 400% tax, not the 80% tax that the British government claimed (BBC, 2000). He saw the relationship between £8 and £10 and he saw the comparison to U.S. sales tax. I discussed this example later with a group of students who could all divide 8 by 10 and 8 by 2, but who could not see where the caller got 400% and 80%. These students knew algorithms, but were lacking in insight.

Acquiring insight is difficult. It involves reflection, judgment, and above all, experience. School curricula seldom emphasize insight as an

*Better known as α and γ.

explicit goal. If asked, most teachers would probably say that insight, if it occurs at all, develops as a by-product of learning mathematical principles. Although many modern curricula do attempt to build insight, there is still no commonly accepted method for doing so. The Third International Mathematics and Science Study (TIMSS) described the curriculum in the United States as "a mile wide, an inch deep" (U.S. National Research Center, 1996). These analysts hope that fewer topics and greater depth will lead to more insight.

But isn't insight required in mathematics? Indeed it is. In fact, insight is frequently what distinguishes a good mathematician from a poor one. But traditionally, the distinction between students with and without insight becomes vital only after calculus. At that point the division between the "haves" and the "have nots" often occurs by natural selection—those who do not have insight drop out of mathematics—rather than by teaching insight. Thus, at the school level, there is no preexisting channel to which we can easily assign the task of teaching insight. A new mechanism is needed. To be effective, the responsibility must be shared by many disciplines.

Quantitative Methods Throughout the Curriculum

My own students describe any shared teaching effort (such as quantitative literacy across the curriculum) as "a conspiracy." A partnership among departments is apparently uncommon enough to look like a conspiracy. A good-natured conspiracy is exactly what we need, however. It is telling that students find it surprising when one field reflects what is being done in another, or when the same ideas come up in several courses. When quantitative literacy is the norm, this will no longer be surprising.

Quantitative literacy is achieved when students readily use quantitative tools to analyze a wide variety of phenomena. This requires constant practice. It also requires seeing such behavior as commonplace. This will not happen unless teachers model it. Verbal literacy became universal when it was perceived to be essential; quantitative literacy will be the same. No matter what we say or what curriculum we teach, students will remain unconvinced of the need for quantitative literacy if they do not perceive their teachers as being quantitatively literate. Even if their teachers are quantitatively literate, we still have an uphill battle because there are many successful citizens who are not; however, ensuring that quantitative literacy permeates the school curriculum is an essential first step.

How do we arrange an infusion of quantitative literacy into the curriculum? Teachers will not simply relinquish time from their own courses; everyone considers his or her field to have been shortchanged already. It cannot be done at the high school level without the involvement of colleges and universities; high schools will not recognize its importance if colleges and universities do not model it. Thus, we need an interdisciplinary partnership that involves high schools, colleges, and universities. Such a partnership must have the backing of business and government, but it cannot be restricted to these communities.

Teachers in this partnership will be asked to take every possible opportunity to encourage students to look at course material through a quantitative lens. Unless they are mathematics teachers, however, it does not mean teaching the quantitative methods themselves. It means teaching students how to identify a quantitative structure and demonstrating the usefulness of a quantitative argument. For mathematics teachers, it means giving more time, perhaps equal time, to developing students' insight in recognizing mathematical ideas in context. For all teachers, it means taking part in an ongoing interdisciplinary dialogue. Such a partnership can lead to a college admission process that rewards quantitative literacy, high-stakes tests that reflect quantitative literacy, and college and high school courses that make frequent and substantial use of quantitative arguments.

Major change does not happen in this country without public conviction that change is necessary. With inspired leadership, a broad-based partnership of policymakers and educators has the potential to convince the public of the importance of quantitative literacy. This is a monumental but vital task.

REFERENCES

Talk of the Nation. Public Broadcasting Service (PBS). September 18, 2000.
"World Fuel Crisis/UK Fuel Tax: The Facts." BBC News, BBC Web page. September 21, 2000.
Summary of Eighth Grade Curriculum Results. U.S. National Research Center, Report No. 7, p. 5. December 1996.

Setting Greater Expectations for Quantitative Learning

Carol Geary Schneider
Association of American Colleges and Universities

To borrow a phrase that has echoed through an era, "The times they are a-changin'."

At the turn of the twentieth century, only about four percent of all Americans went to college. Today, as we enter the twenty-first century, nearly eighty percent of high school students say they would like to go on to higher education and over seventy percent actually do enroll in some form of postsecondary learning within two years of graduating from high school. Many others return to college later in life.

At the turn of the twentieth century, only a few Americans expected to spend their lives engaged in knowledge-based forms of work. Today, creative intelligence is both expected and required at virtually every level of the workplace, from the front desk, to technology-assisted work processes, to environmental analysis and strategic planning. Administrative assistants and executives alike are assessed on their ability to analyze situations, invent appropriate procedures, and solve problems. A healthy percentage of these problems involve numbers.

At the turn of the twentieth century, many key issues were decided, both for the nation and for local communities, by small groups of "leading citizens," i.e., prosperous white men. Our nation's highest court had made "separate but equal" the societal standard and people of color, as well as women, were consistently marginalized. Today, the United States celebrates diversity, and in all quarters of our nation there is a new emphasis on ensuring that every citizen—including those historically excluded—understands and is engaged in the important issues that affect the quality of our lives together. For this expectation to be anything more than a platitude, every citizen, most especially those historically disenfranchised, needs to develop ease and facility in dealing with complex questions, including questions that come framed in terms of quantitative numbers and arguments.

I offer this list of transformational changes in our expectations about the role of advanced knowledge in our world, not to suggest that we have resolved our historical problems with societal asymmetries—we have not—but to set a social and historical context for the very useful "Case for Quantitative Literacy."

The Way We Were

Ours is indeed a world infused with numbers, as the case statement reminds us. But ours is also a world in which a quite new expectation is emerging that "everybody counts" (National Research Council, 1989). This expectation—which has only recently been articulated—challenges deeply entrenched social and institutional practices invented for a world in which everybody was not expected to count, either literally or symbolically. To support high levels of educational accomplishment for everyone, we need to identify and change those dysfunctional practices. As the case statement explains, the school-based mathematics curriculum has been full of such practices. It bears considerable responsibility for the innumeracy that currently characterizes most college graduates.

When I think about my own experience studying mathematics in high school, it is very clear that the course of study I took—the standard lineup of algebra I and II, geometry, trigonometry—was not designed to teach me to deal with quantitatively framed questions relating to the larger society. In my adult life, as a senior academic administrator addressing complex questions about educational and institutional practice, I deal with quantitatively argued issues virtually all the time. But the truth is, there is no connection at all between the mathematics I took beyond arithmetic and the questions I face as a professional. What I understand about quantitative reasoning in my life's work I have picked up on a need-to-know basis, outside of school.

The mathematics curriculum I took in school seemed rather to have two other dominant purposes. The first was sifting. Could students deal sufficiently well with abstract analysis and logical problem solving that they would qualify as "college material"? A diligent student, even when solving problems that seemed pointless to my teenage self, I met that entry-level mathematics standard fully. I did well in the required courses, I crammed with review books to ensure a good SAT score, and I happily became part of the top tier that went on, not just to college, but to a "nationally ranked" college. For most of those who were similarly sifted, whether in or out, mathematics courses were mainly a hurdle to be traversed. They were only incidentally about valuable learning.

The second purpose of the mathematics curriculum seemed to me, even then, to be one of sorting. Did the student's grasp of mathematical principles and practices reach a sufficient level that he or she (mostly he) could expect to perform well in the sciences, or even in advanced mathematics itself? I knew very well that I did not come close to meeting that standard. Science and mathematics would not want me; I accepted that.

Once secure in college, aided and abetted by a flexible system of "distribution requirements," I was careful never again to take a quantitatively oriented course. Mathematicians have told me that my youthful perceptions were close to the mark. The mathematics curriculum has indeed had as one of its major functions selecting the small minority of students who can both embrace and thrive in a world of highly abstract and often lonely analysis.

Looking back on my mathematics experiences, I am reminded of a comment made by a student about a lecture class he had taken with over seven hundred classmates. "If they had wanted us to use what we learned," he observed, "they wouldn't have taught it that way."

The case statement makes the same point. If we really wanted all students to use quantitative strategies in their life, it asserts, we would teach them—students and strategies alike—very differently. We would open the curriculum to often-excluded content, including statistics. And we would use context-attentive pedagogical approaches. For most students, after all, "skills learned free of context are skills devoid of meaning and utility." I was a perfect illustration of this argument.

Were I unique, this little tale would be of no interest whatsoever. But in fact my story is all too typical. The world is infused with quantitative questions, yet the standard mathematics curriculum all too frequently produces mathematics avoiders and amnesiacs.

Setting Greater Expectations

What, then, is to be done? How do we meet the new test of ensuring that an entire nation of college-going citizens develops lasting facility in quantitative reasoning? The case statement offers two primary themes—different content, different pedagogies—as a point of departure. What specific educational changes flow from this prescription?

Here are some proposals that respond to this question. They are drawn from a new Association of American Colleges and Universities (AAC&U) education initiative, entitled "Greater Expectations," that seeks to achieve a new intentionality about what it will take to successfully prepare an entire generation—that new majority now flocking to college—for the intellectual and social demands of the contemporary world.

The Greater Expectations initiative focuses on important outcomes of college-level learning, outcomes that are intended more powerfully to prepare students for lives of creative and thoughtful intelligence, professional excellence, and engaged citizenship. The initiative calls for:

1. Articulation of and focus on forms of learning that are widely needed in the modern world.

2. A new intentionality about addressing expectations for student achievement across successive levels of learning, from school through college.

3. Involvement of students in "authentic assignments," i.e., the kinds of tasks that actually develop complex abilities while showing students how those abilities can be used with power in real contexts.

4. Transparent assessments, linked to authentic assignments, that emphasize what students can do with their knowledge rather than their ability to pass standardized tests.

5. Connection of desired capabilities to learning in each student's major, so that study in the major becomes an essential vehicle not only for developing those capabilities but also for learning how to put them to use.

What do these premises imply for fostering quantitative literacy through school and college learning? Here are my proposals for educational change:

- *Create a public and policy dialogue about the uses of quantitative literacy.*

The first change is to identify, as the authors of the case statement have, the ways in which quantitative literacies are actually used in contemporary society. But this should be more than an academic discussion; the case statement could well be used to spark a broader public and policy dialogue about the need to recast and broaden our expectations for the quantitative literacy of the citizenry.

- *Identify kinds of learning.*

The second change is to move beyond typologies of numeracy to a delineation of the kinds and levels of learning that need to be addressed, both in school and college, if students are actually to be held accountable for developing usable capabilities in quantitative reasoning and problem solving. Here again, the discussions should include policy and civic leaders as well as teachers and scholars.

- *Rethink high school mathematics.*

The third change is to acknowledge the need to substantially retool the high school mathematics curriculum as well as the preparation of the teachers who provide that curriculum. High school study must lay a foundation for statistical as well as mathematical understanding. And it needs to incorporate context-rich practices that enable students to learn essential skills and discover why and for what purpose these skills matter.

- *Rethink college quantitative literacy requirements.*

The fourth change is to recognize that, at the college level, no one course of study can realistically develop all the major kinds of quantitative literacies described in the case statement. We need to stop thinking that remedying our quantitative deficiencies is simply a matter of "fixing" mathematics standards and the corresponding curriculum.

- *Encourage alternative pathways.*

Instead—the fifth change—we need to design multiple courses of study, each well structured to foster quantitative strategies used in specific kinds of professional and civic contexts. The analogy, as the case statement suggests, is to writing. Although all educated people need certain kinds of writing abilities, successful people actually deploy very different rhetorics depending on the context. Scientists, for example, make highly field-specific written arguments; politicians frame their written arguments in very different terms. We should allow college students to develop quantitative strengths keyed to their actual interests, even at the cost of underdeveloping other possible abilities that, realistically, they are unlikely actually to use.

• *Embed quantitative literacy in other fields.*

The sixth change follows from the fifth. It is time to give up on the stand-alone general education mathematics requirement. The great majority of colleges and universities, whether research- or teaching-oriented, still insist that most students take such a course (usually selected from a limited menu of options) as a requirement for graduation. But very little is actually accomplished through this traditional approach to quantitative reasoning and we must fundamentally rethink it. One promising strategy is to make field-related quantitative competence the standard, holding students accountable for evidence of developed ability to actually use quantitative reasoning in ways keyed to their major field(s) of study.

This sixth proposal may give the reader pause. Suppose the student's field of study seems not to require quantitative abilities. What about English, the paradigmatic nonquantitative major?

The tough question is how to bring all fields into dialogue with the modern world. Even as I was majoring in history in the late 1960s, and assiduously avoiding all quantitative courses, my field was actually moving in a decidedly quantitative direction. Most fields, as the case statement reminds us, are becoming more quantitative, reflecting trends in the world at large. All curricula must adapt to these realities. Today many history departments hold students accountable for knowledge of quantitative methods. Tomorrow (or at least in a few years) English departments, already richly infused with sociocultural concerns, must recognize and engage their students' need for quantitative literacy as well.

Moreover, there is a discernible trend on college campuses toward minors and double majors. Colleges might insist that students choose at least one area of concentrated study, whether a major or a minor, that requires and fosters quantitative competence.

Whatever strategy we choose, we must recognize that it really is malpractice to allow students to slip through college without developing the ability to use quantitative strategies to examine significant questions. As the case statement so richly conveys, we are only shortchanging our graduates with respect to the actual demands of a numbers-infused world.

Faculty Work

Lurking beneath all discussions about expectations, curricula, new ways of structuring student learning, and so on are, of course, important concerns about the professional roles of faculty members and about the autonomy and intellectual standing of mathematics departments. No mathematics department wants to see its curriculum cannibalized as each neighboring department incorporates a customized quantitative component. Nor does any mathematics department want to find itself providing only educational "services" to other programs. These are real issues and they cannot be dismissed lightly.

But we are a creative people. Exciting curricular models already abound across the United States in which faculty are linking content courses from different departments together so that students can explore important topics from multiple disciplinary angles. Mathematics is already engaging economics, physics, business, and education. And conversely, most fields are elevating their own expectations for quantitative literacy, raising the possibility of cross-disciplinary collaborations at advanced levels of mathematics rather than in entry-level programs only.

Just as faculty research interests already have blurred disciplinary boundaries, so too curricular innovations can reconfigure the inherited autonomy of departments in intellectually exciting ways. By focusing on contexts, creativity, and new connections across disciplines and fields, scholars who love mathematics may well find new forms of intellectual satisfaction in raising the quantitative literacy of an entire society.

REFERENCES

National Research Council. *Everybody Counts: A Report to the Nation on the Future of Mathematics Education.* Washington, DC: National Academy Press, 1989.

Embracing Numeracy

Lynn Arthur Steen
St. Olaf College

The dawn of the third millennium occasioned unusual public interest in numbers, recalling earlier eras in which numbers were thought to have meanings that influenced or predicted world affairs. Sophisticates dismissed such talk as mere numerology, unbecoming the modern age in which numbers do our bidding, not us theirs. Computers, cell phones, DNA analysis, digital music, and video editing document our mastery of numbers that convey digitally coded instructions. Notwithstanding embarrassing worries about potential Y2K bugs, the high priests and roaring economy of the computer age offered living proof that the new millennium was indeed a new age. It surely seemed, at the turn of the twenty-first century, that by harnessing quantities to amplify our minds we had finally gained full control over the power of numbers.

Yet things are not always as they seem. Consider, for example, Americans' exercise of democracy in the year 2000. It began with political bickering about apportionment and the decennial census—an argument about how to count people. It then moved to the presidential campaign, in which candidates jousted about budgets and tax cuts, about Social Security and medical costs—raising arguments about how to count money. After the election it continued in the courts and the media as partisans contested the result—arguing endlessly about how to count votes.

We might view these events as the revenge of the numbers. It was the number counted ("actual enumeration"), not the people themselves, which would determine apportionment. It was numbers in a spreadsheet—projected surpluses—not actual revenue, which would determine government budget policy. And it was the number of votes tabulated, not the intentions of voters, which would elect the president. Everywhere we looked, numbers seemed once again to be in charge.

Counting people, counting dollars, and counting votes are part of the numeracy of life. Unlike the higher mathematics that is required to design bridges or create cell phones, counting appears to require only rudimentary arithmetic. To be sure, when large numbers, multiple components, and interacting factors are involved, the planning required to ensure accurate counts does become relatively sophisticated. So even though the underlying quantitative concepts are typically rather elementary—primarily topics such as multiplication, percentages, and ratios—the mental effort required to comprehend and solve realistic counting problems is far from simple.

Numeracy and Mathematics

Contrary to popular belief, only a small part of the secret to control over numbers can be found in the mathematics curriculum. That is because skill in complex counting, like dozens of other examples mentioned in "The Case for Quantitative Literacy," is rarely developed in school mathematics. Once basic arithmetic is mastered (or more often, "covered"), the curriculum moves on to advanced and abstract mathematical concepts. Like other rigorous disciplines, mathematics in school advances inexorably toward increasingly sophisticated concepts required for higher education and subsequent professional use. Seldom do students gain parallel experience in applying quantitative skills in subtle and sophisticated contexts.

A chief message of this volume is that more mathematics does not necessarily lead to increased numeracy. Although perhaps counterintuitive, this conclusion follows directly from a simple insight: numeracy is not so much about understanding abstract concepts as about applying elementary tools in sophisticated settings. As the respondents to the case statement emphasize, this is no simple feat. Numeracy takes years of study and experience to achieve. Thus numeracy and mathematics should be complementary aspects of the school curriculum. Both are necessary for life and work, and each strengthens the other. But they are not the same subject.

Respondents to the case statement make this distinction very clear. "I deal with quantitatively argued issues virtually all the time," writes a historian and academic administrator. "But the truth is, there is no connection at all between the mathematics I took beyond arithmetic and the questions I face as a professional." This theme of regret is echoed even by a Ph.D. mathematician: "I was very 'well trained.' Nonetheless, the mathematics education that I received was in many ways impoverished." Another mathematician describes this frustration through the eyes of her students: "What was unexpected was the vehemence with which the students reacted to the scientific context [of a mathematics problem]. . . . They literally wore themselves out by unnecessary efforts." And a mathematics teacher wonders aloud about the motivation of parents: "Why are so many educated people so eager to visit upon their children what might reasonably be considered to be the mistakes of their past?"

The fact that many respondents resonated personally with the case statement's attempt to describe and highlight this new domain called quantitative literacy is itself significant. Each respondent had a similar experience with mathematics in school: canonical courses in the academic track taught with traditional templates and learned well enough to launch professional careers. But as these responses reveal, the challenge of the case statement stimulated in the respondents a sense of something missing, some important preparation for life that was ignored by this traditional mathematics education. The result is a series of uncommonly personal conversations about mathematics and numeracy, about their similarities and differences, their benefits and costs. The stories told in these responses illuminate an important human dimension of mathematics education that is seldom described in public.

Although the mathematics of Euclid and Newton has always held a place of honor in the curriculum, it is only in the last half-century that this particular mathematics came to be regarded as a core subject for everyone anticipating higher education when widespread applications of mathematics made the case for its importance both to individuals and to our nation. The role played by mathematical methods in World War II, the Cold War, the Space Race, and the Computer Age created an unassailable warrant for school mathematics. Apart from English, mathematics is now the most widely taught subject in school and is supported by a cadre of teachers with high aspirations. The nation's mathematics teachers, in fact, were the first to launch a campaign for national standards in education.

The post-Sputnik evolution of mathematics education is as international as mathematics itself. In the early years, the focus was clearly on accelerated learning of mathematics, the goal to move more students more rapidly into the higher reaches of mathematically based fields. Then the pendulum of mathematics education swung back from reform to restoration, from "new math" to "back to basics." Today the pendulum is once again swinging, this time wildly and unpredictably. In some school districts it moves in one direction, in others the opposite. Nations now regularly compare their students' performance in mathematics (and science), and the weak U.S. performance has led many states to impose, for the first time, a mathematics test as a requirement for high school graduation.

For the most part, this reform energy has been devoted to mathematics, not to numeracy. The mainstream mathematics curriculum moves students along the traditional progression from algebra to calculus, not from arithmetic to numeracy. No one has seriously tried to design a school curriculum that gives priority to quantitative literacy as described in the case statement. This is no doubt due in part to the weight of tradition. Mathematics, after all, has a much longer history in schools than does quantitative literacy. In official school tests, in college admissions, and thus in the public mind, it is not numeracy but mathematics that matters for students' futures.

The lack of widespread pressure for quantitative literacy may also be because numeracy is largely invisible to the public. True, the discipline of mathematics is not widely known either. But parents demand mathematics in school because they recognize that it remains today, as it was when they were students, a critical gatekeeper for future opportunities. The ubiquitous need for quantitative thinking is somewhat newer, so there is as yet little public recognition of its pervasive role in daily life and work.

Even the word "numeracy" is relatively new in the American lexicon. Indeed, the first widespread use of the term in the United States was somewhat indirect, being encapsulated in its negation: *Innumeracy*— John Allen Paulos' surprisingly popular outcry against quantitative illiteracy (Paulos, 1988). In other English-speaking nations, (e.g., England, Australia, South Africa) the word "numeracy" is widely used in commentary about education, most often in relation to the early years.

Indeed, governments in these countries have been explicitly pressing a numeracy agenda by urging educators to make school mathematics more socially useful.

In this country, arguments for numeracy come mostly from individuals, not governments. For example, to emphasize the need for effective communication of quantitative information, Yale political science professor Edward Tufte created an extraordinary three-volume work beginning with *The Visual Display of Quantitative Information* on the use and misuse of illustrations to convey numerically based ideas (Tufte, 1983, 1990, 1997). To make the case for the importance of quantitative literacy in contemporary life, financial consultant Peter Bernstein has argued in his best-selling monograph *Against the Gods* that most of modern civilization has been made possible by our ability to understand and control risk (Bernstein, 1996).

Various expressions of numeracy can be found in the bibliography at the end of this volume, as well as some recent expositions of mathematics. These sources not only illustrate the similarities and distinctions between mathematics and numeracy, but they also document the important need for both in students' education. Unfortunately, although every school attempts to deliver a strong mathematics program, vitally important aspects of numeracy such as communication of quantitative information and calibration of risks are all but invisible in standard school curricula.

Although quantitative literacy is a recent and still uncommon addition to the curriculum, its roots in data give it staying power. Mathematics thrived as a discipline and as a school subject because it was (and still is) the tool par excellence for comprehending ideas of the scientific age. Numeracy will thrive similarly because it is the natural tool for comprehending information in the computer age. As variables and equations created the mathematical language of science, so digital data are creating a new language of information technology. Numeracy embodies the capacity to communicate in this new language.

Numeracy Initiatives

As masses of data began to impact ordinary people, as applications of elementary mathematics became pervasive in life and work, and as business and higher education began to question the effectiveness of traditional school mathematics, educators began to explore new approaches to improving students' quantitative experiences and capabilities. Perhaps because colleges operate with fewer encumbrances on curricular change than do secondary schools, most of these initiatives have taken place in postsecondary education.

Indeed, campus initiatives in the area of quantitative literacy are both numerous and diverse. The considerable variety of approaches attests to a healthy climate of experimentation but also to a surprising lack of consensus about either means or ends. In contrast to relatively stable subjects such as Freshman Composition, Introductory French, or Elementary Calculus, colleges seem to have no clear vision about the goals of quantitative literacy or the means by which these goals can most readily be achieved.

Examples of current projects and initiatives whose goals support quantitative literacy include:

Chance. A partnership of several institutions led by Dartmouth College, the Chance program develops courses that study important current news items whose understanding requires a knowledge of chance and risk. These courses are not designed to replace introductory courses in statistics; their goal, rather, is to encourage students to think more rationally about chance events and to make them more informed readers of the daily press. The program includes a professional development component for middle and high school teachers. [*Contact:* Peter Doyle; <doyle@math.dartmouth.edu>]

Exploring Data in Hartford. A mathematics proficiency requirement administered by the Aetna Mathematics Center at Trinity College, this program focuses on skills and concepts in four kinds of relationships (numerical, statistical, algebraic, and logical) that correspond to four proficiency courses: Contemporary Applications: Mathematics for the 21st Century, Cityscape: Analyzing Urban Data, Earth Algebra: Modeling Our Environment, and *Hartford Courant* Issues: Logic in the Media. All courses offered by the Mathematics Center are anchored in contexts using real data from the city of Hartford. [*Contact:* Judith F. Moran; <judith.moran@mail.cc.trincoll.edu>]

Foundations of Scientific Inquiry. A graduation requirement for all arts and science freshmen at New York University with three sequential components: Quantitative Reasoning (Understanding the Mathematical Patterns in Nature), Natural Science I (Introduction to the Physical Universe), and Natural Science II (Our Place in the Biological Realm). For each component students are offered a limited number of choices; for Quantitative Reasoning, the choices are mathematical patterns in nature, mathematical patterns in society, or mathematics and the computer. Quantitative Reasoning is designed to teach students to recognize mathematical patterns within verbally presented problems. The emphasis is neither on technical skills ("Kafkaesque algebraic manipulations") nor on memorization of facts, but on mathematics as the art of pattern recognition. [*Contact:* Fred Greenleaf; <fg3@scivis.nyu.edu>]

Laboratories and Literacy: With the support of grants from the National Science Foundation, the philosophy department at Trinity College (Hartford) has been collaborating with other departments to develop laboratories attached to philosophy courses. Jointly designed by philosophers, scientists, and mathematicians, these laboratories are modeled on those traditionally found in science departments. The laboratories develop specific problem-solving strategies central to mathematics and science, while philosophy classes examine the theoretical accounts and justifications of these strategies. [*Contact:* Helen Lang; <helen.lang@trincoll.edu>]

Mathematics Across the Curriculum (MAC). A National Science Foundation-initiated project at the University of Nevada, Reno, this program is designed to improve students' numeracy skills by helping faculty in different disciplines enhance the quantitative and mathematical content of their courses. One component of MAC is a series of "gateway" examinations that test the mathematical skills students will be required to have for success in various courses in different departments. [*Contact:* Jerry Johnson; <jerryj@unr.edu>]

Mathematical Sciences and their Applications Throughout the Curriculum (MATC). As a sequel to its support of calculus reform, the National Science Foundation awarded grants to several higher education consortia to demonstrate how instruction in the mathematical sciences could be improved by incorporating other disciplinary perspectives. MATC consortia were established at the University of

Pennsylvania, Rensselaer Polytechnic Institute, Dartmouth College, Indiana University, Oklahoma State University, State University of New York at Stony Brook, and the U.S. Military Academy at West Point. [*Information:* <www.matc.siam.org>]

Quantitative Literacy for the Life Sciences. An initiative of the University of Tennessee, Knoxville, this project is designed to enhance the quantitative components of undergraduate life science courses by illustrating the utility of quantitative methods across the full spectrum of the life sciences. Data-based quantitative examples for entry-level biology courses are used to illustrate key biological concepts. [*Contact:* Lou Gross; <gross@math.utk.edu>]

Quantitative Reasoning Across the Curriculum. A faculty development program at Hollins College focused on adapting ideas and materials from successful MATC programs at other institutions. To achieve this goal, Hollins brings project leaders to campus to lead interactive workshops for Hollins faculty who are adapting MATC ideas for quantitative reasoning courses across the curriculum. [*Contact:* Caren Diefenderfer; <cdiefend@hollins.edu>]

Workshop Mathematics. A project begun at Dickinson College, Workshop Mathematics developed three introductory-level courses (Quantitative Reasoning, Statistics, and Calculus with Review) primarily for at-risk students from underserved populations. These courses emphasize active learning, conceptual understanding, real-world applications, and use of technology. They broaden access to university-level mathematics by providing multiple entry points for students who have anxiety about studying mathematics or who do not respond to traditional modes of instruction. [*Contact:* Nancy Baxter Hastings; <baxter@dickinson.edu>]

Next Steps

These projects represent only a tiny fraction of the many quantitative literacy activities already under way. They illustrate college and university efforts because this kind of initiative is more likely to be distinct and visible. Numeracy in secondary schools is harder to detect and describe, primarily because, as an interdisciplinary enterprise, it must live in an atmosphere dominated by pressure for disciplinary standards and, recently, by vigorous arguments about competing mathematics curricula. But here and there we can hear voices urging that school curricula heed the call of quantitative literacy.

To the individuals who helped create "The Case for Quantitative Literacy" we add the support (and occasional caution) of our several respondents. Although the engagement of these educators with numeracy is relatively new, their words and personal stories promise sustained attention. Taken as a whole, this book can clarify the character and importance of numeracy by situating it realistically in an interdisciplinary educational context, and by helping educational leaders understand how the newcomer numeracy relates to the old-timer mathematics.

Based on ideas advanced in this study, the National Council on Education and the Disciplines hopes to make quantitative literacy a priority for schools and colleges across the United States. By creating conversation, supporting initiatives, and encouraging communication among diverse numeracy efforts, NCED aims to shine a spotlight on the need for a different and more effective approach to teaching quantitative skills, an approach rooted in contexts across the curriculum rather than in abstractions that resonate only within mathematics class. In particular, NCED intends to support pilot projects and sites around the country, to sponsor a Web site to support teachers and inform the public, to issue occasional reports and analyses concerning the role of numeracy in contemporary society, and to work with professional organizations to improve articulation of expectations as students move from secondary to higher education.

Numeracy is not the same as mathematics, nor is it an alternative to mathematics. Rather, it is an equal and supporting partner in helping students learn to cope with the quantitative demands of modern society. Whereas mathematics is a well-established discipline, numeracy is necessarily interdisciplinary. Like writing, numeracy must permeate the curriculum. When it does, also like writing, it will enhance students' understanding of all subjects and their capacity to lead informed lives.

REFERENCES

Bernstein, Peter L. *Against the Gods: The Remarkable Story of Risk*. New York, NY: John Wiley, 1996.

Paulos, John Allen. *Innumeracy: Mathematical Illiteracy and its Consequences*. New York, NY: Vintage Books, 1988.

Tufte, Edward R. *The Visual Display of Quantitative Information; Envisioning Information; Visual Explanations—Images and Quantities, Evidence and Narrative* (3 vols.). Cheshire, CT: Graphics Press, 1983, 1990, 1997.

BIBLIOGRAPHY

A selection of readings and references, most quite recent, on numeracy, mathematics, and quantitative literacy.

Alonso, William and Paul Starr. *The Politics of Numbers*. New York, NY: Russell Sage Foundation, 1987.

Banchoff, Thomas F. *Beyond the Third Dimension: Geometry, Computer Graphics, and Higher Dimensions*. New York, NY: Scientific American Library, 1996.

Bennett, Deborah J. *Randomness*. Cambridge, MA: Harvard University Press, 1998.

Berlinski, David. *The Advent of the Algorithm: The Idea That Rules the World*. New York, NY: Harcourt, 2000.

Bunch, Bryan. *The Kingdom of Infinite Number: A Field Guide*. New York, NY: W. H. Freeman, 2000.

Butterworth, Brian. *What Counts: How Every Brain Is Hardwired for Math*. New York, NY: Free Press, 1999.

Caplow, Theodore, Louis Hicks, and Ben J. Wattenberg. *The First Measured Century: An Illustrated Guide to Trends in America, 1900–2000*. Washington, DC: AEI Press, 2001.

Casti, John L. *Five More Golden Rules: Knots, Codes, Chaos, and Other Great Theories of 20th Century Mathematics*. New York, NY: John Wiley, 2000.

Clawson, Calvin C. *Mathematical Sorcery: Revealing the Secrets of Numbers*. New York, NY: Plenum Press, 1999.

Cohen, Patricia Cline. *A Calculating People: The Spread of Numeracy in Early America*. Chicago, IL: University of Chicago Press, 1982; New York, NY: Routledge, 1999.

Cohen, Victor. *News and Numbers*. Ames, IA: Iowa State University Press, 1989.

Cole, K. C. *The Universe and the Teacup: The Mathematics of Truth and Beauty*. New York, NY: Harcourt Brace, 1999.

Crosby, Alfred W. *The Measure of Reality: Quantification and Western Society, 1250–1600*. Cambridge, UK: Cambridge University Press, 1997.

Crump, Thomas. *The Anthropology of Numbers*. Cambridge, UK: Cambridge University Press, 1990.

Dehaene, Stanislas. *The Number Sense: How the Mind Creates Mathematics*. New York, NY: Oxford University Press, 1997.

Desrosières, Alain. *The Politics of Large Numbers: A History of Statistical Reasoning*. Cambridge, MA: Harvard University Press, 1998.

Devlin, Keith J. *Infosense: Turning Data and Information into Knowledge*. New York, NY: W. H. Freeman, 1999.

Devlin, Keith J. *Mathematics: The Science of Patterns*. New York, NY: Scientific American Library, 1994; W. H. Freeman, 1997.

Devlin, Keith J. *The Language of Mathematics: Making the Invisible Visible*. New York, NY: W. H. Freeman, 2000.

Devlin, Keith J. *The Math Gene: How Mathematical Thinking Evolved and Why Numbers Are Like Gossip*. New York, NY: Basic Books, 2000.

Devlin, Keith J. *Goodbye, Descartes: The End of Logic and the Search for a New Cosmology of the Mind*. New York, NY: John Wiley, 1997.

Devlin, Keith J. *Life by the Numbers*. New York, NY: John Wiley, 1998.

Dewdney, A. K. *200% of Nothing: An Eye-Opening Tour Through the Twists and Turns of Math Abuse and Innumeracy*. New York, NY: John Wiley, 1996.

Eastaway, Rob and Jeremy Wyndham. *Why Do Buses Come in Threes? The Hidden Mathematics of Everyday Life*. New York, NY: John Wiley, 1998.

Ekeland, Ivar. *The Broken Dice and Other Mathematical Tales of Chance*. Chicago, IL: University of Chicago Press, 1993.

Gal, Iddo (Editor). *Developing Adult Numeracy: From Theory to Practice*. Cresskill, NJ: Hampton Press, 2000.

Gazalé, Midhat J. *Gnomon From Pharaohs to Fractals*. Princeton, NJ: Princeton University Press, 1999.

Gazalé, Midhat J. *Number: From Ahmes to Cantor*. Princeton, NJ: Princeton University Press, 2000.

Gladwell, Malcolm. *The Tipping Point: How Little Things Can Make a Big Difference*. Boston, MA: Little Brown, 2000.

Herendeen, Robert A. *Ecological Numeracy: Quantitative Analysis of Environmental Issues*. New York, NY: John Wiley, 1998.

Hersh, Reuben. *What Is Mathematics Really?* New York, NY: Oxford University Press, 1997.

Hildebrandt, Stefan and Anthony Tromba. *The Parsimonious Universe: Shape and Form in the Natural World*. New York, NY: Springer-Verlag, 1996.

Hobart, Michael E. and Zachary S. Schiffman. *Information Ages: Literacy, Numeracy, and the Computer Revolution*. Baltimore, MD: Johns Hopkins University Press, 1998.

Hopkins, Nigel J., John W. Mayne, and John W. Hudson. *Go Figure! The Numbers You Need for Everyday Life*. Detroit, MI: Visible Ink Press, 1992.

Huff, Darrell. *The Complete How to Figure It*. New York, NY: W. W. Norton, 1996.

Ifrah, Georges. *The Universal History of Numbers: From Prehistory to the Invention of the Computer*. New York, NY: John Wiley, 1999.

Kaplan, Robert. *The Nothing That Is: A Natural History of Zero*. New York, NY: Oxford University Press, 1999.

Kogelman, Stanley and Barbara R. Heller. *The Only Math Book You'll Ever Need*, Revised Edition. New York, NY: HarperCollins, 1995.

Lakoff, George and Rafael Núñez. *Where Mathematics Comes From: How the Embodied Mind Brings Mathematics into Being*. New York, NY: Basic Books, 2000.

Levin, Simon. *Fragile Dominion: Complexity and the Commons*. Reading, MA: Perseus Books, 1999.

Lord, John. *Sizes: How Big (or Little) Things Really Are*. New York, NY: Harper Perennial, 1995.

MacNeal, Edward. *Mathsemantics: Making Numbers Talk Sense*. New York, NY: Viking Penguin, 1994.

McLeish, John. *The Story of Numbers: How Mathematics Has Shaped Civilization*. New York, NY: Fawcett Columbine, 1994.

Moore, David S. *Statistics: Concepts and Controversies*, Third Edition. New York, NY: W. H. Freeman, 1995.

Morrison, Philip and Phylis Morrison. *Powers of Ten*. New York, NY: Scientific American Books, 1982.

National Council of Teachers of Mathematics. *Principles and Standards for School Mathematics*. Reston, VA: National Council of Teachers of Mathematics, 2000.

National Research Council. *Everybody Counts: A Report to the Nation on the Future of Mathematics Education*. Washington, DC: National Academy Press, 1989.

Noss, Richard and Celia Hoyes. *Windows on Mathematical Meanings*. Dordrecht: Kluwer Academic, 1996.

Nunes, Terezinha, et al. *Street Mathematics and School Mathematics*. New York, NY: Cambridge University Press, 1993.

Organization for Economic Cooperation and Development. *Literacy Skills for the Knowledge Society*. Washington, DC: Organization for Economic Cooperation and Development, 1998.

Orkin, Michael. *What Are the Odds? Chance in Everyday Life*. New York, NY: W. H. Freeman, 1999.

Osserman, Robert. *Poetry of the Universe: The Mathematical Imagination and the Nature of the Cosmos*. New York, NY: Doubleday, 1996.

Paulos, John Allen. *A Mathematician Reads the Newspaper.* New York, NY: Doubleday, 1996.

Paulos, John Allen. *Beyond Numeracy: Ruminations of a Numbers Man.* New York, NY: Alfred A. Knopf, 1991; Vintage Books, 1992.

Paulos, John Allen. *Innumeracy: Mathematical Illiteracy and Its Consequences.* New York, NY: Vintage Books, 1988.

Peterson, Ivars. *The Jungles of Randomness.* New York, NY: John Wiley, 1998.

Poovey, Mary. *History of the Modern Fact: Problems of Knowledge in the Sciences of Wealth and Society.* Chicago, IL: University of Chicago Press, 1998.

Porter, Theodore M. *Trust in Numbers: The Pursuit of Objectivity in Science and Public Life.* Princeton, NJ: Princeton University Press, 1995.

Rima, Ingrid Hahne (Editor). *Measurement, Quantification and Economic Analysis: Numeracy in Economics.* New York, NY: Routledge, 1995.

Rothstein, Edward. *Emblems of Mind: The Inner Life of Music and Mathematics.* New York, NY: Random House, 1995.

Ruelle, David. *Chance and Chaos.* Princeton, NJ: Princeton University Press, 1993.

Schabas, Margaret. *A World Ruled by Number: William Stanley Jevons and the Rise of Mathematical Economics.* Princeton, NJ: Princeton University Press, 1990.

Schoenfeld, A. H. "On Mathematics as Sense-Making: An Informal Attack on the Unfortunate Divorce of Formal and Informal Mathematics." In *Informal Reasoning and Education,* edited by J. Voss, D. Perkins, and J. Segal. Hillsdale, NJ: Erlbaum, 1990, pp. 311–343.

Seife, Charles. *Zero: The Biography of a Dangerous Idea.* New York: Viking Press, 2000.

Shaffner, George. *The Arithmetic of Life.* New York, NY: Ballantine Books, 1999.

Shaffner, George. *Living by the Numbers: Life Lessons From the Bottom Line Up.* New York, NY: Ballantine Books, 2001.

Slavin, Stephen L. *All the Math You'll Ever Need: A Self-Teaching Guide,* Revised Edition. New York, NY: John Wiley, 1999.

Steen, Lynn Arthur (Editor). *Why Numbers Count: Quantitative Literacy for Tomorrow's America.* New York, NY: The College Board, 1997.

Steen, Lynn Arthur. "Numeracy." *Daedalus,* 119:2 (Spring 1990) 211–231.

Steen, Lynn Arthur. "Numeracy: The New Literacy for a Data-Drenched Society." *Educational Leadership,* 57:2 (October 1998) 8–13.

Steen, Lynn Arthur. "Reading, Writing, and Numeracy." *Liberal Education,* 86:2 (Summer 2000) 26–37.

Stigler, Stephen M. *Statistics on the Table: The History of Statistical Concepts and Methods.* Cambridge, MA: Harvard University Press, 1999.

Stutely, Richard. *The Economist Numbers Guide: The Essentials of Business Numeracy.* New York, NY: John Wiley, 1991, 1997.

Swetz, Frank. *Capitalism and Arithmetic: The New Math of the 15th Century*. La Salle, IL: Open Court, 1987.

Tanur, Judith M., et al. *Statistics: A Guide to the Unknown*, Third Edition. Laguna Hills, CA: Wadsworth, 1989.

Traub, J. F. and Arthur G. Werschulz. *Complexity and Information*. New York, NY: Cambridge University Press, 1998.

Tufte, Edward R. *Envisioning Information*. Cheshire, CT: Graphics Press, 1990.

Tufte, Edward R. *The Visual Display of Quantitative Information*. Cheshire, CT: Graphics Press, 1983.

Tufte, Edward R. *Visual Explanations—Images and Quantities, Evidence and Narrative*. Cheshire, CT: Graphics Press, 1997.

Usiskin, Zalman. "From 'Mathematics for Some' to 'Mathematics for All.'" In *Selected Lectures from the 7th International Congress on Mathematical Education,* edited by David F. Robitaille, David H. Wheeler, and Carolyn Kieran. Sainte Foy, Quebec: Les Presses de L'Université Laval, 1994.

Watts, Duncan J. *Small Worlds: The Dynamics of Networks between Order and Randomness*. Princeton, NJ: Princeton University Press, 1999.

White, Stephen. *The New Liberal Arts*. New York, NY: Alfred P. Sloan Foundation, 1981.

Wise, Norton M. *The Values of Precision*. Princeton, NJ: Princeton University Press, 1995.